WIND ENERGY SYSTEMS

Solutions for Power Quality and Stabilization

WIND ENERGY SYSTEMS

Solutions for Power Quality and Stabilization

MOHD. HASAN ALI

CRC Press
Taylor & Francis Group
Boca Raton London New York

CRC Press is an imprint of the
Taylor & Francis Group, an **informa** business

CRC Press
Taylor & Francis Group
6000 Broken Sound Parkway NW, Suite 300
Boca Raton, FL 33487-2742

First issued in paperback 2017

© 2012 by Taylor & Francis Group, LLC
CRC Press is an imprint of Taylor & Francis Group, an Informa business

No claim to original U.S. Government works
Version Date: 20111208

ISBN 13: 978-1-138-07612-9 (pbk)
ISBN 13: 978-1-4398-5614-7 (hbk)

Library of Congress Cataloging-in-Publication Data

Ali, Mohd. Hasan.
  Wind energy systems : solutions for power quality and stabilization / Mohd. Hasan Ali.
    p. cm.
  Includes bibliographical references and index.
  ISBN 978-1-4398-5614-7 (hbk. : alk. paper)
  1. Wind energy conversion systems. 2. Wind power plants. I. Title.

TK1541.A43 2012
621.31'2136--dc23   2011046925

Visit the Taylor & Francis Web site at
http://www.taylorandfrancis.com

and the CRC Press Web site at
http://www.crcpress.com

Contents

Preface

This book provides fundamental concepts of wind energy conversion systems and discusses grid integration and stability issues, methods of transient stability enhancement and minimization of fluctuations of power, and frequency and voltage of wind generator systems. Recently, electricity generation using wind power has received much attention all over the world. Wind energy is a free, renewable resource, so no matter how much is used today there will still be the same supply in the future. Wind energy is also a source of clean, nonpolluting electricity. Unlike conventional power plants, wind plants emit no air pollutants or greenhouse gases.

Induction machines are mostly used as wind generators. However, induction generators have stability problems similar to the transient stability of synchronous machines. During a fault in the power network, rotor speed of the wind generator goes very high, active power output goes very low, and terminal voltage goes low or collapses down. Usually the wind generator is shut down during these emergency situations. Recent tradition is not to shut down the wind generator during a network fault but to keep it connected to the grid through appropriate control. This clearly indicates that wind generator stabilization is necessary during network faults. Again, even though there is no fault in the network, due to random wind speed variations, output power, frequency, and terminal voltage of wind generators fluctuate. However, consumers desire to have constant voltage and frequency. Therefore, some control means are necessary to minimize power, frequency, and voltage fluctuations. This book discusses several means to enhance the transient stability of wind generator system and also explains the methodologies for minimizing fluctuations of power, frequency, and voltage.

The book is organized as follows. Chapter 1 provides a general overview of wind energy and outlines the background, aim, and scope of the book. Chapter 2 describes the fundamental concept of wind energy conversion systems and modeling of wind turbines. Electric machines—in particular induction machines and synchronous machines—are the key to wind energy conversion systems. Therefore, the basic concepts of electrical machines are discussed in Chapter 3. With the variable-speed wind generator system, the terminal of the generator is connected to the grid through a power electronics interface. Also, to integrate the energy storage system into the wind generator system, power electronics is necessary. Thus, Chapter 4 deals with a brief overview and fundamental concepts of power electronics devices. The types of wind generator systems are discussed in Chapter 5.

Chapter 6 discusses the grid integration issues of wind generator systems. In particular, the transient stability issue, power quality problem, fluctuations of power, frequency, and voltage of wind generator systems during random wind speed variations are described. Chapter 7 analyzes the solutions for power quality issues of wind generator systems, especially minimization of fluctuations of power, voltage, and frequency of wind generators during random wind speed variations by energy storage devices. Chapter 8 describes the various methods of transient stability enhancement of wind generator systems during network faults. Simulation results are provided to demonstrate the effectiveness of the stabilization methods. Comparisons among the stabilization methods are made on the basis of performance, control structure, and cost. Chapter 9 deals with the fault-ride through capability and mitigation of power fluctuations of variable-speed wind generator systems, especially for doubly fed induction generator systems, wound-field synchronous generator systems, and permanent magnet synchronous generator systems.

This book is intended to discuss various means of wind generator stabilization and the means to minimize power, frequency, and voltage fluctuations of wind generator systems. A comparison of stabilization tools is also given; thus, it will help the researchers and engineers understand the relative effectiveness of the stabilization methods and then select a suitable tool for wind generator stabilization. This book can also serve as a good reference for wind generator systems for graduate as well as undergraduate students. Through this book, these students will gain insight into grid integration and stability issues and various methods of stabilization of wind generator systems and can apply the knowledge they gain from this book in their own research. Thus, it is hoped this book will be of great interest and be very helpful for researchers, engineers, and students who research wind energy systems.

The author cordially invites any constructive criticism of or comments about the book.

Acknowledgments

I would like to express my sincere gratitude to all those who devoted their help to my completion of the book. I must thank the staff of CRC Press, LLC, especially Nora Konopka, Kathryn Younce, Joselyn Banks-Kyle, and Jennifer Ahringer, whose guidance was invaluable in preparing my first book manuscript. My thanks also go to my wife, Shammi Akhter, and my daughter, Aniqa Tahsin Ali, who patiently stood by me with deep understanding and continuous support while I was preoccupied with the project.

Mohd. Hasan Ali
University of Memphis
Memphis, Tennessee

About the Author

Mohd. Hasan Ali received a BScEng in electrical and electronic engineering from Rajshahi University of Engineering and Technology (RUET), Rajshahi, Bangladesh, in 1995, and an MScEng and PhD in electrical and electronic engineering from Kitami Institute of Technology, Kitami, Japan, in 2001 and 2004, respectively. He was a lecturer from 1995 to 2004 and assistant professor in 2004 in the Department of Electrical and Electronic Engineering at RUET. He was a postdoctoral research fellow at the Kitami Institute of Technology, Japan, under the Japan Society for the Promotion of Science (JSPS) Program from November 2004 to January 2007. He was also a research professor with the Department of Electrical Engineering at the Changwon National University, South Korea, during 2007. Ali served as postdoctoral research fellow with the Department of Electrical and Computer Engineering at Ryerson University, Canada, from 2008 to 2009. From 2009 to 2011 he was a faculty member with the Department of Electrical Engineering at the University of South Carolina. Currently, he is an assistant professor at the electrical and computer engineering department of the University of Memphis, Tennessee.

Ali's research interests include power system dynamics, stability and control, smart-grid and micro-grid power systems, renewable energy systems, energy storage systems, electric machines and motor drives, application of power electronics to power systems, flexible alternating current transmission systems (FACTS) devices, and application of fuzzy logic control to power systems. He has authored and coauthored more than 100 publications, including journal and conference papers. Ali is a senior member of the IEEE Power and Energy Society.

chapter 1

Overview

The terms *wind energy* and *wind power* describe the process by which the wind is used to generate mechanical power or electricity. Wind turbines convert the kinetic energy in the wind into mechanical power. This mechanical power can be used for specific tasks (e.g., grinding grain or pumping water), or a generator can convert this mechanical power into electricity to power homes, businesses, schools, and the like. Recently, generation of electricity using wind power has received much attention all over the world. This chapter provides an overview of wind energy systems.

1.1   Introduction

Wind energy is a free, renewable resource, so no matter how much is used today there will still be the same supply in the future. Wind energy is also a source of clean, nonpolluting electricity. Unlike conventional power plants, wind plants emit no air pollutants or greenhouse gases. Currently, extensive research on wind energy is going on in various countries of the world, including the United States, Germany, Spain, Denmark, Japan, South Korea, Canada, Australia, and India. There are several organizations for wind energy research in the world, like the Global Wind Energy Council (GWEC), National Renewable Energy Laboratory (NREL), and American Wind Energy Association (AWEA). According to a GWEC report [1], about 12% of the world's total electricity demand can be supplied by the wind energy by 2020. This figure indicates the importance of wind energy research these days.

1.2   Why Renewable Energy

Coal, oil, and gas, which are being used as fuels for conventional power plants, are being depleted gradually, so exploration of alternative fuel sources—that is, renewable energy sources for producing electricity—is needed. There are various types of renewable energy sources, like solar energy, wind energy, geothermal, and biomass. Renewable energy is the use of nonconventional energy sources to generate electrical power and fuel vehicles for today's residential, commercial, institutional, and industrial energy applications. This includes emergency power systems, transportation systems, on-site electricity generation, uninterrupted power supply, combined heat and power systems, off-grid power systems, electrical peak-shaving systems, and many more innovative applications.

Wind power has a great advantage over conventional fuels. Its operation does not produce harmful emissions or any hazardous waste. It does not deplete natural resources in the way that fossil fuels do, nor does it cause environmental damage through resource extraction, transport, and waste management.

The generation of electricity by wind turbines is dependent on the strength of the wind at any given moment. Wind farm sites are chosen after careful analysis to determine the pattern of the wind—its relative strength and direction at different times of the day and year. So wind power is variable but not unpredictable.

1.3   Wind Energy

Wind power or wind energy is the process by which the wind is used to generate mechanical power or electrical power and is one of the fastest-growing forms of electrical power generation in the world. The power of the wind has been used for at least 3,000 years. Until the twentieth century, wind power was used to provide mechanical power to pump water or to grind grain. At the beginning of modern industrialization, the use of the fluctuating wind energy resource was substituted by fossil-fuel-fired engines or the electrical grid, which provided a more consistent power source. So the use of wind energy is divided into two parts: (1) mechanical power generation; and (2) electrical power generation.

Wind is simple air in motion. It is caused by the uneven heating of the earth's surface by the sun. Since the earth's surface is made of very different types of land and water, it absorbs the sun's heat at different rates. During the day, the air above the land heats up more quickly than the air over water. The warm air over the land expands and rises, and the heavier, cooler air rushes in to take its place, creating winds. At night, the winds are reversed because the air cools more rapidly over land than over water. In the same way, the large atmospheric winds that circle the earth are created because the land near the earth's equator is heated more by the sun than the land near the North and South Poles.

Today, wind energy is mainly used to generate electricity. Wind is called a renewable energy source because the wind will blow as long as the sun shines.

1.4   Advantages and Disadvantages of Wind-Generated Electricity

1.4.1   A Renewable Nonpolluting Resource

Wind energy is a clean, reliable cost effective source of electricity. Electricity generated from the wind does not contribute to global warming and acid rain. Compared to energy from nuclear power plants, there is no risk of radioactive exposure from wind power.

1.4.2   Cost Issues

Even though the cost of wind power has decreased dramatically in the past 10 years, the technology requires a **higher initial investment** than fossil-fueled generators. Roughly 80% of the cost is the machinery, with the balance being site preparation and installation. If wind-generating systems are compared with fossil-fueled systems on a "life-cycle" cost basis (counting fuel and operating expenses for the life of the generator), however, wind costs are much more competitive with other generating technologies because there is no fuel to purchase and minimal operating expenses.

1.4.3   Environmental Concerns

Although wind power plants have relatively little impact on the environment compared with fossil fuel power plants, there is some concern over the **noise** produced by the rotor blades, **aesthetic (visual) impact**, and the death of birds and bats caused from flying into the rotors. Most of these problems have been resolved or greatly reduced through technological development or by properly siting wind plants.

1.4.4   Supply and Transport Issues

The major challenge to using wind as a source of power is that it is **intermittent** and does not always blow when electricity is needed. Wind cannot be stored (although wind-generated electricity can be stored if batteries are used), and not all winds can be harnessed to meet the timing of electricity demands. Further, good wind sites are often located in **remote locations** far from areas of electric power demand (such as cities). Finally, wind resource development may compete with other uses for the land, and those **alternative uses** may be more highly valued than electricity generation. However, wind turbines can be located on land that is also used for grazing or even farming.

1.5   Worldwide Status of Wind Energy

The following provides an overview of the worldwide status of wind energy based on the GWEC report [1].

1.5.1   Europe

Wind is the fastest growing power technology in Europe. Although Europe was home to only one-third of the world's new installed capacity in 2008, the European market continues its steady growth, and wind power is now the fastest growing power generation technology in the European Union. Indeed, 43% of all new energy installations in 2008 was wind power, well ahead of gas (35%) and oil (13%).

Overall, almost 8.9 GW of new wind turbines brought European wind power generation capacity up to nearly 66 GW. There is now clear diversification of the European market, relying less and less on the traditional wind markets of Germany, Spain, and Denmark. The year 2008 saw a much more balanced expansion, with a "second wave" led by Italy, France, and the United Kingdom. Of the EU's 27 member states, 10 now have more than 1 GW of wind power capacity.

In 2008 the European wind turbine market was worth €11 billion. The entire wind fleet will produce 142 TWh of electricity, or about 4.2% of EU demand, in an average wind year. This will save about 100 m tons of CO2 each year.

1.5.2   Germany

Though at the global level Germany has been surpassed by the United States, it continues to be Europe's leading market, in terms of both new and total installed capacity. Over 1.6 GW of new capacity was installed in 2008, bringing the total up to nearly 24 GW. Wind energy is continuing to play an important role in Germany's energy mix. In 2008, 40.4 TWh of wind power was generated, representing 7.5% of the country's net electricity consumption. In economic terms, too, wind power has become a serious player in Germany, and the sector now employs close to 100,000 people.

1.5.3   Spain

Spain is Europe's second largest market and has seen growth in line with previous years (with the exception of 2007, when regulatory change brought about a higher than usual amount of new wind capacity). In 2008, 1.6 GW of new generating equipment was added to the Spanish wind fleet, bringing the total up to 16.7 GW. This development confirms Spain as a steadily growing market, which at this rate is likely to reach the government's 2010 target of 20 GW of installed wind capacity.

In 2008, wind energy generated more than 31,000 GWh, covering more than 11% of the country's electricity demand.

1.5.4 Italy

One noteworthy newcomer among the growing European markets in 2008 was Italy, which experienced a significant leap in wind power capacity. Over 1,000 MW of new wind turbines came online in 2008, bringing the total installed capacity up to 3.7 GW.

1.5.5 France

France is also continuing to see strong growth, after progressing steadily in recent years. In 2000, France had only 30 MW of wind-generating capacity, mostly small wind turbines in the French overseas territories. At the end of 2008, the total installed capacity stood at 3.4 GW, representing an annual growth rate of 38%.

Wind power is now France's fastest growing energy source; in 2008, around 60% of all new power generation capacity in France was wind energy. The biggest potential in the coming years is estimated to be in the north and the northeast of the country. Out of 4,000 MW of approved wind power projects, more than 700 MW are in the region Champagne-Ardennes and 500 MW in Picardy.

1.5.6 United Kingdom

Despite being host to some of the best wind resources in Europe, the United Kingdom's wind energy market has taken a long time to start realizing this potential. For the first time, in 2009 the UK wind energy sector delivered more than 1,000 MW of new wind power capacity in 1 year, spread over 39 wind farms. From January to December 2009 a total of 1,077 MW of capacity was installed on and offshore, taking the United Kingdom's installed total to 4,051 MW.

1.5.7 European Union

The European Union continues to be the world's leader in total installed wind energy capacity and one of the strongest regions for new development, with over 10 GW of new installed capacity in 2009. Industry statistics compiled by the European Wind Energy Association (EWEA) show that cumulative EU wind capacity increased by 15% to reach a level of 74,767 MW, up from 64,719 MW at the end of 2008.

In the European Union, wind power is by far the most popular electricity-generating technology. For the second year running, wind energy had the largest market share: Of almost 26 GW installed in the European Union in 2009, wind power accounted for 39%. All renewable technologies combined accounted for 61% of new power-generating capacity. Since 2000, installed wind capacity has increased from 9.7 GW to 75 GW.

By the end of the year, a total of more than 190,000 people were employed in the wind energy sector, and investments in wind farms amounted to about €13 billion in the European Union during 2009. The wind power capacity installed by the end of 2009 will, in a normal wind year, produce 162.5 TWh of electricity, equal to 4.8% of the European Union's electricity demand. Spain and Germany remain the two largest annual markets for wind power, competing each year for the top spot (2,459 MW and 1,917 MW, respectively, in 2009), followed by Italy (1,114 MW), France (1,088 MW), and the United Kingdom (1,077 MW). A total of 11 EU member states—over one-third of all EU countries—now each has more than 1,000 MW of installed wind energy capacity. Austria and Greece are just below the 1,000 MW mark.

Offshore wind is setting itself up to become a mainstream energy source in its own right. In 2009, 582 MW of offshore wind was installed in the European Union, up 56% from the previous year. Cumulative capacity increased to 2,063 MW. The main markets were the United Kingdom and Denmark. For 2011, it is expected that a further 1,000 MW of offshore wind will be installed in Europe. This would represent around 10% of the 2010 market.

1.5.8   North America

1.5.8.1   United States

In 2008 the United States was the number one market in terms of both new capacity and total wind generation capacity, and it broke all previous records with new installations of 8.5 GW, reaching a total installed capacity of over 25 GW. The massive growth in the U.S. wind market in 2008 doubled the country's total wind-power-generating capacity. The new wind projects completed in 2008 accounted for about 42% of the entire new power producing capacity added in the United States in 2007 and created 35,000 new jobs, bringing the total employed in the sector up to 85,000.

1.5.8.2   Canada

Canada in 2008 surpassed the 2 GW mark for installed wind energy capacity, ending the year with 2.4 GW. Canada's wind farms now produce enough power to meet almost 1% of Canada's total electricity demand.

The year 2008 was Canada's second best ever for new wind energy installations, with 10 new wind farms coming online, representing 526 MW of installed wind energy capacity. Included in this total were the first wind farms in the provinces of New Brunswick, Newfoundland, and Labrador. In British Columbia, the only Canadian province without a wind farm, construction began on the first wind farm, with completion expected in early 2009.

1.5.9 Asia

The growth in Asian markets has been breathtaking, and nearly a third of the 27 GW of new wind energy capacity added globally in 2008 was installed in Asia.

1.5.9.1 China

China continued its spectacular growth in 2008, once again doubling its installed capacity by adding about 6.3 GW, to reach a total of 12.2 GW. The prospects for future growth in the Chinese market are very good. In response to the financial crisis, the Chinese government has identified the development of wind energy as one of the key economic growth areas, and in 2009 new installed capacity is expected to nearly double again. At this rate, China is on its way to overtaking Germany and Spain to reach second place in terms of total wind power capacity in 2011. This means that China would have met its 2020 target of 30 GW 10 years ahead of time.

The growing wind power market in China has also encouraged domestic production of wind turbines and components, and the Chinese manufacturing industry is becoming increasingly mature, stretching over the whole supply chain. According to the Chinese Renewable Energy Industry Association (CREIA), the supply is starting to not only satisfy domestic demand but also meet international needs.

1.5.9.2 India

India is continuing its steady growth, with 1,800 MW of wind energy capacity added in 2008, bringing the total up to 9.6 GW. The leading wind producing state in India is Tamil Nadu, which hosts over 4 GW of installed capacity, followed by Maharashtra with 1.8 GW and Gujarat with 1.4 GW.

1.5.9.3 Japan

Japan's wind energy industry has surged forward in recent years. Development has also been encouraged by the introduction of market incentives, both in terms of the price paid for the output from renewable plants and in the form of capital grants towards clean energy projects. Power purchase agreements for renewables also have a relatively long life span of 15 to 17 years, which helps to encourage investor confidence. The result has been an increase in Japan's installed capacity from 136 MW at the end of 2000 to 1,880 MW at the end of 2008. In 2008, 346 MW of new wind capacity was added in Japan. Other Asian countries with new capacity additions in 2008 include Taiwan (81 MW for a total of 358 MW) and South Korea (43 MW for a total of 236 MW).

1.5.10   Pacific

1.5.10.1   Australia

After several years of stagnation in Australia's wind market, the speed of development picked up again in 2008, with 482 MW of new installations, a 58% leap in terms of total installed capacity. Australia is now home to 50 wind farms, with a total capacity of 1.3 GW. Seven additional projects totaling 613 MW are currently under construction and expected to become operational in 2009.

1.5.10.2   New Zealand

New Zealand's installed capacity grew by only 3.5 MW in 2008 to reach 325 MW. However, this does not adequately reflect the wind industry's activity over the year, and a further 187 MW is currently under construction.

1.5.11   Latin America

The Latin American market, despite the tremendous wind resources in the region, saw only slow growth in 2008.

1.5.11.1   Brazil

The only country installing substantial new capacity is Brazil, which added 94 MW of wind energy across five wind farms, mostly located in Ceará in the northeast of the country. Brazil's Programme of Incentives for Alternative Electricity Sources (PROINFA) was initially passed in 2002 to stimulate the addition of over 1,400 MW of wind energy capacity and other renewable sources. The first stage was supposed to finish in 2008, but it has now been extended.

Mexico offers significant wind energy potential with conditions that are considered among the best in Latin America, especially in the area of La Ventosa, Oaxaca (7,000 MW), with average capacity factors above 40%. In addition, other sites in several states also offer good wind power potential, in particular La Rumorosa in Baja California and sites in Tamaulipas, Yucatán, Veracruz, Zacatecas, Hidalgo, and Sinaloa, Mexico. The Ministry of Energy (SENER) has estimated that a total of 40,000 MW of wind energy potential could be developed in Mexico.

Despite this significant potential, wind development in Mexico has been slow, mainly due to inadequate financial incentives, issues with the existing regulatory framework, and a lack of policies to encourage use of wind energy. However, in 2009, new laws and regulations were introduced to boost renewable energy development, and more than 560 MW of wind projects are currently under construction.

1.5.11.2   Mexican Power Generation System

Mexico has around 50 GW of total installed electricity generation capacity, including 11,457 MW from independent power producers (IPP) and about 7,900 MW of self-generation and cogeneration capacity. During the last years, the generation mix in Mexico changed significantly, moving away from fuel-oil generation plants to natural-gas-based ones. Traditionally, large hydroelectric and geothermal energy have been Mexico's most widely used renewable sources. Other renewable energy sources, such as wind power, photovoltaic (PV), small hydro, biomass, and biofuels have experienced only slow growth to date.

1.5.12   Africa and Middle East

1.5.12.1   130 MW Installed in Africa and Middle East

In North Africa, the expansion of wind power continues in Egypt, Morocco, and Tunisia, with 55 MW, 10 MW, and 34 MW of new capacity installed, respectively. In the Middle East, Iran installed 17 MW of new capacity. The total installed wind energy capacity in Africa and the Middle East now stands at 669 MW.

Figures 1.1, 1.2, 1.3, and 1.4 show the global cumulative installed wind capacity since 1996 to 2010, the total installed capacity in 2008, top 10 cumulative capacity in December 2010, and annual market forecast by region in 2010–2015, respectively [1].

1.6   Aim and Scope of the Book

This book aims to provide basic concepts of wind energy conversion system. Various means to enhance transient stability and to minimize power, frequency, and voltage fluctuations of wind generator systems are discussed. Comparison is made among the stabilization tools to help researchers and engineers understand their relative effectiveness and then to select the best one. Graduate as well as undergraduate students can use this book as a good reference for wind generator systems and can gain insight into grid integration and stability issues and various methods of stabilization of wind generator systems. They can then apply the knowledge they gain from this book in their own research. Thus, it is hoped that this book would be of high interest and very helpful for researchers, engineers, and students intending to perform research in wind energy systems.

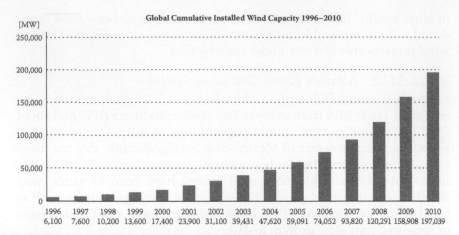

Figure 1.1 (See color insert.) Cumulative installed capacity, 1996–2010.

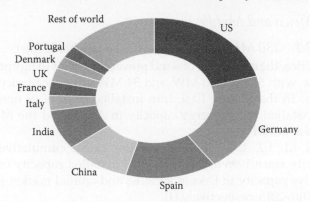

	MW	%
USA	25,170	20.8
Germany	23,903	19.8
Spain	16,754	13.9
China	12,210	10.1
India	9,615	8.0
Italy	3,736	3.1
France	3,404	2.8
UK	3,241	2.7
Denmark	3,180	2.6
Portugal	2,862	2.4
Rest of world	16,693	13.8
Total top 10	**104,104**	**85.2**
World total	**120,798**	**100.0**

Figure 1.2 (See color insert.) Total installed capacity 2008.

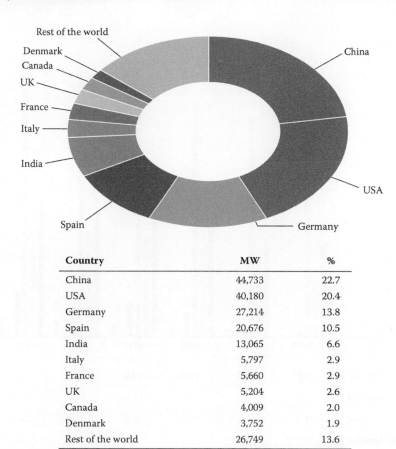

Country	MW	%
China	44,733	22.7
USA	40,180	20.4
Germany	27,214	13.8
Spain	20,676	10.5
India	13,065	6.6
Italy	5,797	2.9
France	5,660	2.9
UK	5,204	2.6
Canada	4,009	2.0
Denmark	3,752	1.9
Rest of the world	26,749	13.6
Total top 10	**170,290**	**86.4**
World total	**197,039**	**100.0**

Figure 1.3 (See color insert.) Top 10 cumulative capacity, December 2010.

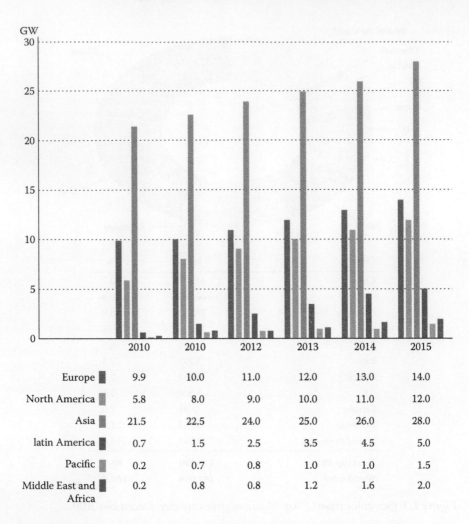

	2010	2010	2012	2013	2014	2015
Europe	9.9	10.0	11.0	12.0	13.0	14.0
North America	5.8	8.0	9.0	10.0	11.0	12.0
Asia	21.5	22.5	24.0	25.0	26.0	28.0
latin America	0.7	1.5	2.5	3.5	4.5	5.0
Pacific	0.2	0.7	0.8	1.0	1.0	1.5
Middle East and Africa	0.2	0.8	0.8	1.2	1.6	2.0

Figure 1.4 (See color insert.) Annual market forecast by region, 2010–2015.

References

1. Global Wind Energy Council (GWEC), http://www.gwec.net/
2. S. Heier, Grid integration of wind energy conversion system, John Wiley & Sons, 1998.
3. A. Sumper, O. G. Bellmunt, A. S. Andreu, R. V. Robles, and J. R. Duran, "Response of fixed speed wind turbines to system frequency disturbances," *IEEE Trans. Power Systems*, vol. 24, no. 1, pp. 181–192, February 2009.
4. Y. Amirat, M. E. H. Benbouzid, B. Bensaker, R. Wamkeue, and H. Mangel, "The state of the art of generators for wind energy conversion systems," *Proceedings of the International Conference on Electrical Machine (ICEM'06)*, pp. 1–6, Chania, Greece, 2006.
5. E. Koutroulis and K. Kalaitzakis, "Design of a maximum power point tracking system for wind-energy-conversion-applications," *IEEE Transactions on Industrial Electronics*, vol. 53, no. 2, pp. 486–494, April 2006.
6. M. H. Ali and B. Wu, "Comparison of stabilization methods for fixed-speed wind generator systems," *IEEE Transactions on Power Delivery*, vol. 25, no. 1, pp. 323–331, January 2010.
7. B. S. Borowy and Z. M. Salameh, "Dynamic response of a stand-alone wind energy conversion system with battery storage to a wind gust," *IEEE Trans. Energy Conversion*, vol. 12, no. 1, pp. 73–78, Mar. 1997.
8. R. Chedid and S. Rahman, "Unit sizing and control of hybrid wind-solar power systems," *IEEE Trans. Energy Conversion*, vol. 12, no. 1, pp. 79–85, Mar. 1997.
9. A. C. Saramourtsis, A. G. Bakirtzis, P. S. Dokopoulos, and E. S. Gavanidou, "Probabilistic evaluation of the performance of wind-diesel energy systems," *IEEE Trans. Energy Conversion*, vol. 9, no. 4, pp. 743–752, December 1994.
10. I. Abouzahr and R. Ramakumar, "An approach to assess the performance of utility-interactive wind electric conversion systems," *IEEE Trans. Energy Conversion*, vol. 6, no. 4, pp. 627–638, December 1991.
11. A. G. Bakirtzis, P. S. Dokopoulos, E. S. Gavanidou, and M. A. Ketselides, "A probabilistic costing method for the evaluation of the performance of grid-connected wind arrays," *IEEE Trans. Energy Conversion*, vol. 4, no. 1, pp. 34–40, March 1989.
12. V. Valtchev, A. Bossche, J. Ghijselen, and J. Melkebeek, "Autonomous renewable energy conversion system," *Renewable Energy*, vol. 19, no. 1, pp. 259–275, January 2000.
13. E. Muljadi and C. P. Butterfield, "Pitch-controlled variable-speed wind turbine generation," *IEEE Trans. Ind. Appl.*, vol. 37, no. 1, pp. 240–246, Jan. 2001.
14. L. H. Hansen et al., "Generators and power electronics technology for wind turbines," *in Proceedings of IEEE IECON'01*, vol. 3, pp. 2000–2005, Denver, CO, November–December 2001.
15. T. Ackermann et al., "Wind energy technology and current status: A review," *Renewable and Sustainable Energy Reviews*, vol. 4, pp. 315-374, 2000.
16. R. W. Thresher et al., "Trends in the evolution of wind turbine generator configurations and systems," *Int. J. Wind Energy*, vol 1, no. 1, pp. 70–86, April 1998.
17. P. Carlin et al., "The history and state of the art of variable-speed wind turbine technology," *Int. J. Wind Energy*, vol. 6, no. 2, pp. 129–159, April–June 2003.

18. A. Grauers et al., "Efficiency of three wind energy generator systems," *IEEE Trans. Energy Conversion*, vol. 11, no. 3, pp. 650-657, September 1996.

19. B. Blaabjerg et al., "Power electronics as efficient interface in dispersed power generation systems," *IEEE Trans. Power Electronics*, vol. 19, no. 5, pp. 1184-1194, September 2004.

20. P. Thoegersen et al., "Adjustable speed drives in the next decade. Future steps in industry and academia," *Electric Power Components & Systems*, vol. 32, no. 1, pp. 13–31, January 2004.

21. J. A. Baroudi et al., "A review of power converter topologies for wind generators," in *Proceedings of IEEE IEMDC'05*, pp. 458–465, San Antonio, TX, May 2005.

22. B. Blaabjerg et al., "Power electronics as an enabling technology for renewable energy integration," *J. Power Electronics*, vol. 3, no. 2, pp. 81–89, April 2003.

23. C. Nicolas et al., "Guidelines for the design and control of electrical generator systems for new grid connected wind turbine generators," in *Proceedings of IEEE IECON'02*, vol. 4, pp. 3317–3325, Seville, Spain, November 2002.

24. M. A. Khan et al., "On adapting a small pm wind generator for a multi-blade, high solidity wind turbine," *IEEE Trans. Energy Conversions*, vol. 20, no. 3, pp. 685–692, September 2005.

25. J. R. Bumby et al., "Axial-flux permanent-magnet air-cored generator for small-scale wind turbines," *IEE Proc. Electric Power Applications*, vol. 152, no. 5, pp. 1065–1075, September 2005.

26. G. K. Singh, "Self-excited induction generator research—A survey," *Electric Power Systems Research*, vol. 69, pp. 107–114, 2004.

27. R. C. Bansal et al., "Bibliography on the application of induction generators in nonconventional energy systems," *IEEE Trans. Energy Conversion*, vol. 18, no. 3, pp. 433–439, September 2003.

28. P. K. S. Khan et al., "Three-phase induction generators: A discussion on performance," *Electric Machines & Power Systems*, vol. 27, no. 8, pp. 813–832, August 1999.

29. M. Ermis et al., "Various induction generator schemes for wind-electricity generation," *Electric Power Systems Research*, vol. 23, no. 1, pp. 71–83, 1992.

30. S. Muller et al., "Doubly fed induction generator systems for wind turbines" *IEEE Industry Applications Magazine*, vol. 8, no. 3, pp. 26–33, May–June 2002.

31. R. Datta et al., "Variable-speed wind power generation using doubly fed wound rotor induction machine—A comparison with alternative scheme," *IEEE Trans. Energy Conversion*, vol. 17, no. 3, pp. 414–421, September 2002.

32. L. Holdsworth et al., "Comparison of fixed speed and doubly-fed induction wind turbines during power system disturbances," *IEE Proc. Generation, Transmission and Distribution*, vol. 150, no. 3, pp. 343–352, May 2003.

33. S. Grabic et al., "A comparison and trade-offs between induction generator control options for variable speed wind turbine applications," in *Proceedings of IEEE ICIT'04*, vol. 1, pp. 564–568, Hammamet, Tunisia, December 2004.

34. P. Mutschler et al., "Comparison of wind turbines regarding their energy generation," in *Proceedings of IEEE PESC'02*, vol. 1, pp. 6–11, Cairns, Australia, June 2002.

35. R. Hoffmann et al., "The influence of control strategies on the energy capture of wind turbines," in *Proceedings of IEEE IAS'02*, vol. 2, pp. 886–893, Rome, Italy, October 2000.

36. M. Orabi et al., "Efficient performances of induction generator for wind energy," in *Proceedings of IEEE IECON'04*, vol. 1, pp. 838–843, Busan, Korea, November 2004.

37. J. G. Slootweg et al., "Inside wind turbines—Fixed vs. variable speed," *Renewable Energy World*, pp. 30–40, 2003.

38. P. M. Anderson and A. Bose, "Stability simulation of wind turbine systems," *IEEE Trans. Power Apparatus and Systems*, vol. PAS-102, no. 12, pp. 3791–3795, December 1983.

39. G. L. Johnson, Wind Energy Systems Electronic Edition, http://www.rpc.com.au/products/windturbines/wind book/WindTOC.html

40. NEDO LAWEPS, http://www2.infoc.nedo.go.jp/nedo/top.html

41. NEDO, The New Energy and Industrial Technology Development Organization, http://www.nedo.go.jp/engli sh/introducing/what.html

42. E. S. Abdin and W. Xu, "Control design and dynamic performance analysis of a wind turbine-induction generation unit," *IEEE Trans. Energy Conversion*, vol. 15, no. 1, pp. 91–96, March 2000.

43. J. R. Winkelman and S. H. Javid, "Control design and performance analysis of a 6 MW wind turbine generator," *IEEE Trans. on PAS*, vol. 102, no. 5, pp. 1340–1347, May 1983.

44. O. Wasynczuk, D. T. Man, and J. P. Sullivan, "Dynamic behavior of a class of wind turbine generators during random wind fluctuations," *IEEE Trans. on PAS*, vol. 100, no. 6, pp. 2837–2845, June 1981.

45. F. P. de Mello, J. W. Feltes, L. N. Hannett, and J. C. White, "Application of induction generators in power system," *IEEE Trans. on PAS*, vol. 101, no. 9, pp. 3385–3393, 1982.

46. M. A. Rahman, A. M. Osheiba, T. S. Radwan, and E. S. Abdin, "Modelling and controller design of an isolated diesel engine permanent magnet synchronous generator," *IEEE Trans. on EC*, vol. 11, no. 2, pp. 324–330, June 1996.

47. L. Soder et al., "Experience from wind integration in some high penetration areas," *IEEE Trans. Energy Conversion*, vol. 22, no. 1, pp. 4–12, Mar. 2007.

48. A. Radunskaya, R. Williamson, and R. Yinger, "A dynamic analysis of the stability of a network of induction generators," *IEEE Trans. Power Syst.*, vol. 23, no. 2, pp. 657–663, May 2008.

49. J. Craig, "Dynamics of wind generators on electric utility network," *IEEE Trans. Aerosp Electron. Syst.*, vol. 12, pp. 483–493, July 1976.

chapter 2

Wind Energy Conversion System

2.1 Introduction

Among renewable energy sources, wind energy generation has been noted as the most rapidly growing technology because it is one of the most cost-effective and environmental friendly means to generate electricity from renewable sources. Wind energy conversion systems convert the kinetic energy of the wind into electricity or other forms of energy. Wind power generation has experienced a tremendous growth in the past decade and has been recognized as an environmentally friendly and economically competitive means of electric power generation. This chapter describes the basics of the wind energy conversion system, its fundamental concept, basic components of a wind energy system, and system modeling and types of wind turbines.

2.2 Fundamental Concept

Every wind energy system transforms the kinetic energy of the wind into mechanical or electrical energy. There are huge variations in size, but all wind turbines—from the smallest to the largest—work in the same way. The overall configuration is identical. Each system consists of a rotor (blades) that converts the wind's energy into rotational shaft energy, a nacelle (enclosure) containing a drive train, and a generator [1]. The energy that moves the wind (kinetic energy) moves the blades. This energy in turn moves the drive train (mechanical energy). This energy is then turned into electricity (electrical energy) in the generators and then stored in batteries or transferred to home power grids or utility companies for use in the usual way. Figure 2.1 shows a brief flow diagram of the wind energy conversion system.

2.3 Wind Energy Technology

The first commercial wind turbines were deployed in the 1980s, and since then their installed capacity, efficiency, and visual design have all improved enormously [2]. Although many different pathways toward the ideal turbine design have been explored, significant consolidation has taken place over the past decade. The vast majority of commercial turbines now operate on a horizontal axis with three evenly spaced blades.

Available wind ⟶ Turbine ⟶ Mechanical ⟶ Electrical
      Energy   power   power

Figure 2.1 Brief flow diagram of wind energy conversion system.

These are attached to a rotor from which power is transferred through a gearbox to a generator. The gearbox and generator are contained within a housing called a nacelle. Some turbine designs avoid a gearbox by using direct drive. The electricity is then transmitted down the tower to a transformer and eventually into the grid network.

Wind turbines can operate across a wide range of wind speeds—from 3 to 4 m/s up to about 25 m/s, which translates into 90 km/h (56 mph), which is the equivalent of gale force 9 or 10. The majority of current turbine models make best use of the constant variations in the wind by changing the angle of the blades through *pitch control*, which occurs by turning or *yawing* the entire rotor as wind direction shifts and by operating at variable speed. Operation at variable speed enables the turbine to adapt to varying wind speeds and increases its ability to harmonize with the operation of the electricity grid. Sophisticated control systems enable fine-tuning of the turbine's performance and electricity output.

Modern wind technology is able to operate effectively at a wide range of sites—with low and high wind speeds—in the desert, and in freezing arctic climates. Clusters of turbines collected into wind farms operate with high availability, are generally well integrated with the environment, and are accepted by the public. Using lightweight materials to reduce their bulk, modern turbine designs are sleek, streamlined, and elegant.

The main design drivers for current wind technology are reliability, grid compatibility, acoustic performance (noise reduction), maximum efficiency and aerodynamic performance, high productivity for low wind speeds, and offshore expansion. Wind turbines have also grown larger and taller. The generators in the largest modern turbines are 100 times the size of those in 1980. Over the same period, their rotor diameters have increased eightfold. The average capacity of turbines installed around the world during 2007 was 1,492 kW, whereas the largest turbine currently in operation is the Enercon E126, with a rotor diameter of 126 meters and a power capacity of 6 MW.

The main driver for larger capacity machines has been the offshore market, where placing turbines on the seabed demands the optimum use of each foundation. Fixing large foundations in the seabed, collecting the electricity, and transmitting it to the shore all increase the costs of offshore development over those on land. Although the offshore wind farms installed so far have used turbines in the capacity range up to 3.6 MW, a range of designs of 5 MW and above are now being deployed and are expected to become the standard in the coming years.

For turbines used on land, however, the past few years have seen a leveling of turbine size in the 1.5 to 3 MW range. This has enabled series production of many thousands of turbines of the same design, enabling teething problems to be ironed out and increasing reliability.

Ongoing innovations in turbine design include the use of different combinations of composite materials to manufacture blades, especially to ensure that their weight is kept to a minimum, variations in the drive train system to reduce loads and increase reliability, and improved control systems, partly to ensure better compatibility with the grid network.

2.4 Basic Components of a Wind Turbine System

Typical wind turbines involve a set of rotor blades (usually three) rotating around a hub. The hub is connected to a gearbox and a generator, located inside the nacelle, which houses the electrical components. The basic components of a wind turbine system are shown in Figure 2.2 and outlined as follows:

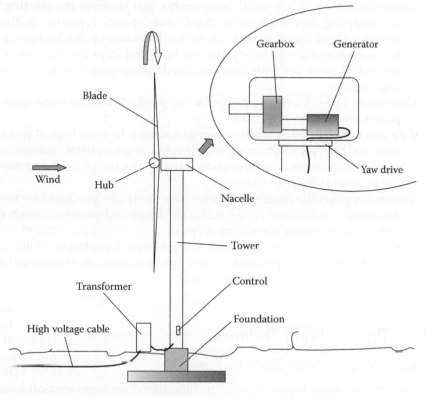

Figure 2.2 Major turbine components.

The nacelle: Sits on top of the tower and contains the electrical components—the gearbox, the brake, the wind speed and director monitor, the yaw mechanism, and the generator.

Rotor blades: The diameter of the blades is a crucial element in the turbine power; typically, the longer they are, the greater the output. But their design and the materials incorporated by them are also key elements. Blades are often made of fiberglass reinforced with polyester or wood epoxy. Vacuum resin infusion is a new material connected to a technology presented by manufacturers like Suzlon. Typically blades rotate at 10–30 revolutions per minute, either at a constant speed (the more traditional solution) or at a variable speed.

Gearboxes and direct drives: Most wind turbines use gearboxes, whose function is to increase the rotational speed required by generators. Some new technologies are exploring direct drives generators to dispense with the expensive gears.

Brake: A disk used to stop the rotor blades in emergencies and to ensure the safety of the turbine in case of very high damaging winds or other exceptional situations.

Controller: A set of electrical components that controls the starting, the stopping, and the turbine rotor blade speed. Typically, in the constant wind speed model the controller starts up the turbine at wind speeds around 8 to 14 miles per hour and stops the machine at around 55 miles per hour (to avoid the damage caused by turbulent high winds).

Generator: The device responsible for the production of 60-cycle alternating current (AC) electricity.

The yaw mechanism of wind power generators: In more typical wind turbines, the yaw mechanism is connected to sensors (e.g., anemometers) that monitor wind direction, turning the tower head and lining up the blades with the wind.

Tower: Supports the nacelle and rotor. The electricity produced by the generator comes down cables inside the tower and passes through a transformer into the electricity network.

Base: Large turbines are built on a concrete base foundation. When a wind turbine ceases production, it is a simple task to dig these out or cover them, leaving little trace behind.

2.5   Types of Wind Turbines

2.5.1   Wind Turbines Based on Axes

There are two main types of wind turbine based on axes: vertical axis wind turbine (VAWT); and horizontal axis wind turbine (HAWT).

Figure 2.3 (See color insert.) Vertical axis wind turbine.

2.5.1.1   *Vertical Axis Wind Turbines*

In the VAWT the generator shaft is positioned vertically with the blades pointing up with the generator mounted on the ground or a short tower. Figure 2.3 shows a conventional vertical axis wind turbine. There are two basic types of airfoils (blades): lift and drag:

1. Drag-type VAWT: In drag-type, the blades are generally a flat plate that the wind hits and causes to rotate. This type of design is great for very low-wind areas and will develop a lot of torque to perform an operation. However, in medium to higher winds, their capabilities to produce energy are limited.

2. Lift-type VAWT: The lifting style airfoil is seen in most modern wind turbines. A properly designed airfoil is capable of converting significantly more power in medium and higher winds. Actually, with this design, the fewer number of blades, the more efficient this design can be. Each blade sees maximum lift (torque) only twice per revolution, making for a huge torque and sinusoidal power output. Two European companies actually produced a one-bladed machine, but it is not commercially used because of dynamic balance issues.

The basic theoretical advantages of a vertical axis wind turbine are as follows:

1. The generator, gearbox, and so forth may be placed on the ground, and a tower may not be needed for the machine.
2. A yaw mechanism is not needed to turn the rotor against the wind.

The basic disadvantages are as follows:

1. Wind speeds are very low close to ground level, so although a tower may be saved wind speeds will be very low on the lower part of the rotor.
2. The overall efficiency of the vertical axis turbines is not impressive.

2.5.1.2   *Horizontal Axis Wind Turbines*

An HAWT rotates around a horizontal axis and has the main rotor shaft and electrical generator mounted at the top of a tower. Figure 2.4 shows a conventional horizontal axis wind turbine. There are two types of HAWT:

1. Horizontal upwind: The generator shaft is positioned horizontally, and the wind hits the blade before the tower. Turbine blades are made stiff to prevent the blades from being pushed into the tower by high winds, and the blades are placed at a considerable distance in front of the tower and are sometimes tilted up a small amount.
2. Horizontal downwind: The generator shaft is positioned horizontally, and the wind hits the tower first and then the blade. Horizontal downwind doesn't need an additional mechanism for keeping it in line with the wind, and in high winds the blades can be allowed to bend, which reduces their swept area and thus their wind resistance. Horizontal downwind turbine is also free of turbulence problems.

Figure 2.4 (See color insert.) Horizontal axis wind turbine.

2.5.2 Wind Turbine Power Scales

According to the size of the wind turbine and the annual mean wind speed, there are three types of wind turbines [1]:

1. Small-scale wind turbines: Their power range is 0.025 kW to 10 kW. Annual mean wind speed range of 2.5 to 4.0 m/s is needed for this type of wind turbine.
2. Medium-scale wind turbines: Need 4.0 to 5.0 m/s annual wind speed to produce power. Output power range is 10 kW to 100 kW.

3. Large-scale wind turbines: Annual mean wind speed needed is more than 5 m/s, and output power is greater than 100 kW. They are usually connected to grids.

2.5.3   Wind Turbine Installation Location

Depending on installation locations, there are two types of wind turbines: the onshore wind farm and the offshore wind farm, as shown in Figures 2.5 and 2.6, respectively. There are several advantages of the offshore turbines: (1) higher and more constant wind speeds; and, consequently, (2) higher efficiencies. Onshore wind farms are often subject to restrictions and objections: objections based on their negative visual impact or noise; restrictions associated with obstructions (e.g., buildings, mountains), land-use disputes, or limited availability of lands. However, onshore wind systems may also have some advantages over offshore wind farms:

- Cheaper foundations
- Cheaper integration with the electrical-grid network
- Cheaper installation and access during the construction phase
- Cheaper and easier access for operation and maintenance

Figure 2.5 (See color insert.) Onshore wind farm.

Figure 2.6 (See color insert.) Offshore wind turbine.

2.6   Modeling of Wind Turbines

2.6.1   Power Output from an Ideal Turbine

Wind energy is the kinetic energy of the moving air mass. A wind turbine converts the kinetic energy of moving air into mechanical motion that can be either used directly to run the machine or to run the generator to produce electricity.

The kinetic energy, U, in a parcel of air of mass, m, flowing at speed, v, in the x direction is

$$U = \frac{1}{2}mv^2 = \frac{1}{2}(\rho A x)v^2 \qquad Joules \qquad (2.1)$$

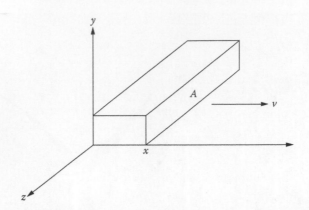

Figure 2.7 Packet of air moving with speed, v.

where A is the cross sectional area in m², ρ is the air density [kg/m³], x is the thickness of the parcel in m. If we visualize the parcel as in Figure 2.7 with side x moving with speed u and the opposite side fixed at the origin, we see the kinetic energy increasing uniformly with x, because the mass is increasing uniformly.

The power in the wind, Pw, is the time derivative of the kinetic energy

$$P_w = \frac{dU}{dt} = \frac{1}{2}\rho Av^2 \frac{dx}{dt} = \frac{1}{2}\rho Av^3 \qquad W \qquad (2.2)$$

This can be viewed as the power being supplied at the origin to cause the energy of the parcel to increase according to equation (2.1). A wind turbine will extract power from side x with equation (2.2) representing the total power available at this surface for possible extraction.

The physical presence of a wind turbine in a large moving air mass modifies the local air speed and pressure as shown in Figure 2.8. The picture is drawn for a conventional horizontal axis propeller type turbine.

Let us consider a tube of moving air with initial or undisturbed diameter, $d1$, speed, $u1$, and pressure, $p1$, as it approaches the turbine. The speed of the air decreases as the turbine is approached, causing the tube of air to enlarge to the turbine diameter, $d2$. The air pressure will rise to a maximum just in front of the turbine and will drop below atmospheric pressure behind the turbine. Part of the kinetic energy in the air is converted to potential energy to produce this increase in pressure. Still more kinetic energy will be converted to potential energy after the turbine to raise the air pressure back to atmosphere. This causes the wind speed to continue to decrease until the pressure is in equilibrium. Once the low point of wind speed is reached, the speed of the tube of air will increase back to $u4$ = $u1$ as it receives kinetic energy from the surrounding air.

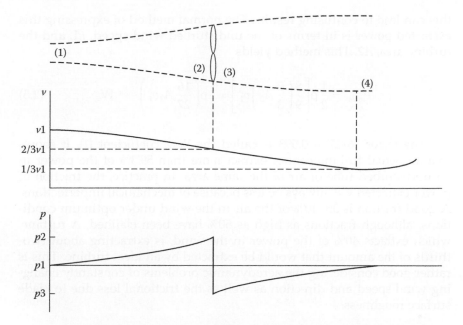

Figure 2.8 Circular tube of air flowing through ideal wind turbine.

It can be shown that under optimum conditions, when maximum power is being transferred from the tube of air to the turbine, the following relationship holds:

$$v_2 = v_3 = \frac{2}{3}v_1$$

$$v_4 = \frac{1}{3}v_1$$

$$A_2 = A_3 = \frac{3}{2}A_1 \qquad (2.3)$$

$$A_4 = 3A_1$$

The mechanical power extracted is then the difference between the input and output power in the wind:

$$P_{mideal} = P_1 - P_4 = \frac{1}{2}\rho\left(A_1 v_1^3 - A_4 v_4^3\right) = \frac{1}{2}\rho\left(\frac{8}{9}A_1 v_1^3\right) \qquad W \qquad (2.4)$$

This states that 8/9 of the power in the original tube of air is extracted by an ideal turbine. This tube is smaller than the turbine, however, and this can lead to confusing results. The normal method of expressing this extracted power is in terms of the undisturbed wind speed, $u1$, and the turbine area, $A2$. This method yields

$$P_{mideal} = \frac{1}{2}\rho\left[\frac{8}{9}\left(\frac{2}{3}A_2\right)v_1^3\right] = \frac{1}{2}\rho\left(\frac{16}{27}A_2v_1^3\right) \quad W \qquad (2.5)$$

The factor $16/27 = 0.593$ is called the Betz coefficient [1]. It shows that an actual turbine cannot extract more than 59.3% of the power in an undisturbed tube of air of the same area. In practice, the fraction of power extracted will always be less because of mechanical imperfections. A good fraction is 35–40% of the air in the wind under optimum conditions, although fractions as high as 50% have been claimed. A turbine, which extracts 40% of the power in the wind, is extracting about two-thirds of the amount that would be extracted by an ideal turbine. This is rather good considering the aerodynamic problems of constantly changing wind speed and direction as well as the frictional loss due to blade surface roughness.

2.6.2   Power Output from Practical Turbines

The fraction of power extracted from the power in the wind by a practical wind turbine is usually given by the symbol Cp, standing for the coefficient of performance or power coefficient. Using this notation and dropping the subscripts of Equation (2.5), the actual mechanical power output can be written as

$$P_m = C_p\left(\frac{1}{2}\rho Av^3\right) = C_pP_m \quad W \qquad (2.6)$$

where P_w is the extracted power from the wind, ρ is the air density [kg/m³], R is the blade radius [m], V_w is the wind velocity [m/s], and C_p is the power coefficient, which is a function of both tip speed ratio, λ, and blade pitch angle, β [deg].

The coefficient of performance is not constant but varies with the wind speed, the rotational speed of the turbine, and turbine blade parameters like angle of attack and pitch angle. Generally it is said that the power coefficient, Cp, is a function of tip speed ratio, l, and blade pitch angle, b, where tip speed ratio is defined as

$$\lambda = \frac{R\omega}{v} \qquad (2.7)$$

where R is the radius of the turbine blade in m, w is the mechanical angular velocity of the turbine in rad/s, and v is the wind speed in m/s. The angular velocity, w, is determined from the rotational speed, n (r/min), by

$$\omega = \frac{2\pi n}{60} \quad rad/s \tag{2.8}$$

Usually, the MOD-2 model is considered for C_p-λ characteristics [3,4], which is represented by the following equations and shown in Figure 2.9. Data for a large 2.5 MW (MOD-2) wind turbine are given in Table 2.1.

$$C_P = \frac{1}{2}\left(\lambda - 0.022\beta^2 - 5.6\right)e^{-0.17\lambda} \tag{2.9}$$

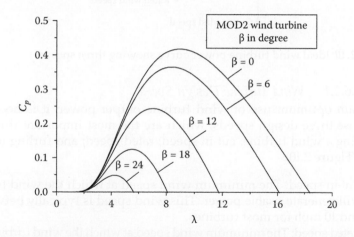

Figure 2.9 C_p-λ curves for different pitch angle.

Table 2.1 2.5 MW (MOD-2) Wind Turbine Data

Turbine Type	Three-Blade Horizontal Axis
Radius	46 m
Gear ratio	1:103
Rotor speed	18 rpm
Air density	1.225 kg/m³
Cut in wind speed	4 m/sec
Rated wind speed	approximately 12 m/sec
Tower height	about 100 m

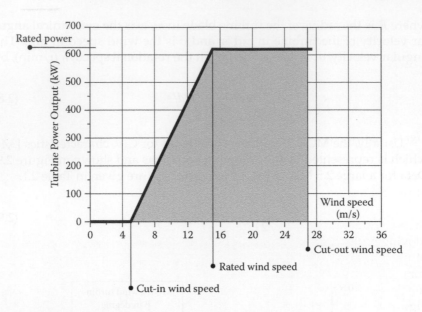

Figure 2.10 Ideal wind turbine power curve showing three speeds.

2.6.2.1   Wind Turbine Design Speed

To obtain optimum use of wind turbine output power, it is necessary to choose three design speeds, which are the most important things in designing a wind turbine: cut-in speed; rated speed; and furling (cutoff) speed (Figure 2.10):

1. Cut-in speed: The minimum wind speed at which the wind turbine will generate usable power. This wind speed is typically between 7 and 10 mph for most turbines.
2. Rated speed: The minimum wind speed at which the wind turbine will generate its designated rated power. Rated speed for most machines is in the range of 25 to 35 mph. At wind speeds between cut-in and rated, the power output from a wind turbine increases as the wind increases. The output of most machines levels off above the rated speed.
3. Furling (cutout) speed: At very high wind speeds, typically between 45 and 80 mph, most wind turbines cease power generation and shut down. The wind speed at which shutdown occurs is called the cut-out speed or sometimes the furling speed. Having a cut-out speed is a safety feature that protects the wind turbine from damage.

2.6.2.2   Pitch Mechanism

Controlling the power output from the turbines blades is a major issue in any wind turbine, and it can be accomplished by two main technologies.

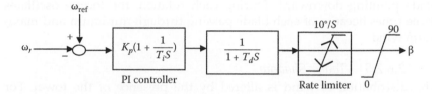

Figure 2.11 Pitch control system model.

One, known as the pitch control technology, is based on the adjustment of the blades by a control system, intimately connected to the brake system. The other technology, the stall (or passive) control, is purely based on the aerodynamic properties of the blade, without any control system or moving parts to adjust. It's a technology based on the twist and thickness of the blade along its length. Usually the main purpose of using a pitch controller with a wind turbine is to maintain a constant output power at the terminal of the generator when the wind speed is over the rated speed. Figure 2.11 shows the conventional pitch control system [5].

2.6.2.3   Effect of Wind Shear and Tower Shadow

Torque and power generated by a wind turbine are much more variable than that produced by more conventional generators. The sources of these power fluctuations are due both to stochastic processes that determine the wind speeds at different times and heights and to periodic processes. These periodic processes are largely due to two effects termed *wind shear* and *tower shadow* [6]. Wind shear is used to describe the variation of wind speed with height, whereas tower shadow describes the redirection of wind due to the tower structure. In three-bladed turbines, the most common and largest periodic power pulsations occur at what is known as a 3p frequency. This is three times the rotor frequency, or the same frequency at which the blades pass by the tower. Thus, even for a constant wind speed at a particular height a turbine blade would encounter variable wind as it rotates. Torque pulsations, and therefore power pulsations, are observed due to the periodic variations of wind speed experienced at different locations.

2.6.2.4   Wind Shear

Wind speed generally increases with height, and this variation is wind shear. Torque pulsations, and therefore power pulsations, are observed due to the periodic variations of wind speed seen at different heights. Power and torque oscillate due to the different wind conditions encountered by each blade as it rotates through a complete cycle. For instance, a blade pointing upward would encounter wind speeds greater than a blade pointing downward. During each rotation, the torque oscillates three times because of each blade passing through minimum and maximum wind.

2.6.2.5   Tower Shadow

The distribution of wind is altered by the presence of the tower. For upwind rotors, the wind directly in front of the tower is redirected and thereby reduces the torque at each blade when in front of the tower. This effect is called tower shadow. The torque pulsations due to tower shadow are most significant when a turbine has blades downwind of the tower and wind is blocked as opposed to redirected. For this reason, the majority of modern wind turbines have upwind rotors.

2.7   Chapter Summary

This chapter provides the fundamental concept of wind energy conversion system. The wind energy technology, the basic components of a wind turbine system, types of wind turbines are described. Wind turbine power scales and installation location are discussed. The equations of output power considering ideal and practical turbines are derived. The pitch control mechanism, wind shear and tower shadow phenomena are explained.

References

1. S. Heier, *Grid integration of wind energy conversion system,* John Wiley & Sons, 1998.
2. Global Wind Energy Council (GWEC), http://www.gwec.net/
3. P. M. Anderson and A. Bose, "Stability simulation of wind turbine systems," *IEEE Trans. Power Apparatus and Systems,* vol. PAS-102, no. 12, pp. 3791–3795, December 1983.
4. J. G. Slootweg, S. W. D. de Haan, H. Polinder, and W. L. Kling, "General model for representing variable speed wind turbines in power system dynamics simulations," *IEEE Trans. Power Systems,* vol. 18, no. 1, pp. 144–151, February 2003.
5. M. H. Ali and B. Wu, "Comparison of stabilization methods for fixed-speed wind generator systems," *IEEE Trans. on Power Delivery,* vol. 25, no. 1, pp. 323–331, January 2010.
6. D. S. L. Dolan and P. W. Lehn, "Simulation model of wind turbine 3p torque oscillations due to wind shear and tower shadow," *IEEE Trans. Energy Conversion,* vol. 21, no. 3, pp. 717–724, September 2006.
7. T. Thiringer and J.-A. Dahlberg, "Periodic pulsations from a three-bladed wind turbine," *IEEE Trans. Energy Conversion,* vol. 16, pp. 128–133, June 2001.
8. T. Thiringer, "Power quality measurements performed on a low-voltage grid equipped with two wind turbines," *IEEE Trans. Energy Conversion,* vol. 11, pp. 601–606, September 1996.

9. E. N. Hinrichsen and P. J. Nolan, "Dynamics and stability of wind turbine generators," *IEEE Trans. Power App. Syst.*, vol. 101, pp. 2640–2648, August 1982.
10. T. Petru and T. Thiringer, "Modeling of wind turbines for power system studies," *IEEE Trans. Power Syst.*, vol. 17, no. 4, pp. 1132–1139, November 2002.
11. J. B. Ekanayake, L. Holdsworth, W. XueGuang, and N. Jenkins, "Dynamic modeling of doubly fed induction generator wind turbines," *IEEE Trans. Power Syst.*, vol. 18, no. 2, pp. 803–809, May 2003.
12. D. J. Trudnowski, A. Gentile, J. M. Khan, and E. M. Petritz, "Fixed speed wind-generator and wind-park modeling for transient stability studies," *IEEE Trans. Power Syst.*, vol. 19, no. 4, pp. 1911–1917, November 2003.
13. J. Cidras and A. E. Feijoo, "A linear dynamic model for asynchronous wind turbines with mechanical fluctuations," *IEEE Trans. Power Syst.*, vol. 17, no. 3, pp. 681–687, August 2002.
14. K. T. Fung, R. L. Scheffler, and J. Stolpe, "Wind energy—A utility perspective," *IEEE Trans. Power App. Syst.*, vol. 100, pp. 1176–1182, 1981.
15. S. H. Karaki, B. A. Salim, and R. B. Chedid, "Probabilistic model of a towsite wind energy conversion system," *IEEE Trans. Energy Conversion*, vol. 17, no. 4, pp. 530–542, 2002.
16. A. E. Feijoo, J. Cidras, and J. L. G. Dornelas, "Wind speed simulation in wind farms for security assessment of electrical power systems," *IEEE Trans. Energy Conversion*, vol. 14, pp. 1582–1587, 1999.
17. P. Poggi, M. Muselli, G. Notton, C. Cristofari, and A. Louche, "Forecasting and simulating wind speed in Corsica by using an autoregressive model," *Energy Conversion and Management*, vol. 44, pp. 3177–3196, 2003.
18. R. Karki and R. Billinton, "Reliability/cost implications of PV and wind energy utilization in small isolated power systems," *IEEE Trans. Energy Conversion*, vol. 16, no. 4, pp. 368–372, 2001.
19. P. Wang and R. Billinton, "Reliability benefit analysis of adding WTG to a distribution system," *IEEE Trans. Energy Conversion*, vol. 16, no. 2, pp. 134–139, 2001.
20. A. Saramourtsis, P. Dokopoulos, A. Bakirtziz, and E. Gavanidou, "Probabilistic evaluation of the performance of wind diesel energy systems," *IEEE Trans. Energy Conversion*, vol. 9, no. 4, pp. 743–752, 1994.
21. P. S. Dokopoulos, A. C. Saramourtsis, and A. G. Bakirtziz, "Prediction and evaluation of the performance of wind diesel energy systems," *IEEE Trans. Energy Conversion*, vol. 11, no. 2, pp. 385–393, 1996.
22. R. Billinton and G. Bai, "Generating capacity adequacy associated with wind energy," *IEEE Trans. Energy Conversion*, vol. 19, no. 3, pp. 641–646, 2004.
23. R. Billinton, H. Chen, and R. Ghajar, "A sequential simulation technique for adequacy evaluation of generating systems including wind energy," *IEEE Trans. Energy Conversion*, vol. 11, no. 4, pp. 728–734, 1996.
24. G. N. Kariniotakis, G. S. Stavrakakis, and E. F. Nogaret, "Wind power forecasting using advanced neural networks models," *IEEE Trans. Energy Conversion*, vol. 11, no. 4, pp. 762–768, 1996.
25. S. Kelouwani and K. Agbossou, "Non-linear model identification of wind turbine with a neural network," *IEEE Trans. Energy Conversion*, vol 19, no. 3, p. 608, 2004.
26. H. Madads, H. M. Kajabadi et al., "Development of a novel wind turbine simulator for wind energy conversion systems using an inverter—controlled induction motor," *IEEE Trans. Energy Conversion*, vol. 19, no. 3, pp. 547–560, 2004.

27. A. G. Bakirtz, "A probabilistic method for the evaluation of the reliability of stand-alone wind energy conversion systems," *IEEE Trans. Energy Conversion*, vol. 7, no. 1, pp. 530–536, 1992.

28. C. Singh and A. Lago-Gonzaloz, "Reliability modeling of generation systems including unconventional energy sources," *IEEE Trans. Power Appl. Syst.*, vol. PAS-104, pp. 1049–1055, 1985.

29. H. Mohammed and C. O. Nwankpa, "Stochastic analysis and simulation of grid-connected wind energy conversion system," *IEEE Trans. Energy Conversion*, vol. 15, no. 1, pp. 85–89, 2005.

30. S. H. Karaki, R. B. Chedid, and R. Ramadan, "Probabilistic performance assessment of wind energy conversion systems," *IEEE Trans. Energy Conversion*, vol. 14, no. 2, pp. 217–224, 1999.

31. R. Chedid and S. Rahman, "Unit sizing and control of hybrid wind solar power systems," *IEEE Trans. Energy Conversion*, vol. 12, no. 1, pp. 79–85, 1997.

32. T. Thiringer and J.-A. Dahlberg, "Periodic pulsation from a three-bladed wind turbine," *IEEE Trans. Energy Conversion*, vol. 16, no. 2, pp. 128–133, 2001.

33. B. I. Daqiang, X. Wang, W. Weijian, and D. Howe, "Improved transient simulation of salient pole synchronous generators with internal and ground faults in the stator winding," *IEEE Trans. Energy Conversion*, vol. 20, no. 1, pp. 128–134, 2005.

34. E. Muljadi, H. L. Hess, and K. Thomas, "Zero sequence method for energy recovery from a variable-speed wind turbine generator," *IEEE Trans. Energy Conversion*, vol. 16, no. 1, pp. 99–103, 2001.

35. S. P. Singh, S. K. Jain, and J. Sharma, "Voltage regulation optimization of compensated self-excited induction generator with dynamic load," *IEEE Trans. Energy Conversion*, vol. 19, no. 4, pp. 724–732, 2004.

36. B. J. Chalmers, E. Spooner, and W. Wu, "An axial-flux permanent-magnet generator for a gearless WES," *IEEE Trans. Energy Conversion*, vol. 14, no. 2, pp. 251–257, 1999.

37. M. Popescu, T. J. E. Miller, N. Trivillin, and O. Robert, "Asynchronous performance analysis of a single-phase capacitor-start, capacitor-run permanent magnet motor," *IEEE Trans. Energy Conversion*, vol. 20, no. 1, pp. 142–149, 2005.

38. P. Alan, W. Patric, and L. Chapman, "Simple expression for optimal current waveform for permanent-magnet synchronous machine drives," *IEEE Trans. Energy Conversion*, vol. 20, no. 1, pp. 151–157, 2005.

39. H. Douglas, P. Pillay, and A. K. Ziarani, "Broken rotor bar detection in induction machines with transient operating speeds," *IEEE Trans. Energy Conversion*, vol. 20, no. 1, pp. 135–141, 2005.

40. A. Siddique, G. S. Yadava, and B. Singh, "A review of stator fault monitoring techniques of induction motors," *IEEE Trans. Energy Conversion*, vol. 20, no. 1, pp. 106–113, 2005.

41. C. Chris, G. R. Slemon, and R. Bonert, "Minimization of iron losses of permanent magnet synchronous machines," *IEEE Trans. Energy Conversion*, vol. 20, no. 1, pp. 121–127, 2005.

42. Z. Chen and E. Spooner, "Grid power quality with variable speed wind turbine," *IEEE Trans. Energy Conversion*, vol. 16, no. 2, pp. 148–153, 2001.

43. F. D. Kanellos and N. D. Hatziargyrious, "The effect of variable-speed wind turbines on the operation of weak distribution networks," *IEEE Trans. Energy Conversion*, vol. 17, no. 4, pp. 543–548, 2002.

44. T. Thiringer, "Power quality measurements performed on a low-voltage grid equipped with two wind turbines," *IEEE Trans. Energy Conversion*, vol. 1, no. 3, pp. 601–606, 1996.

45. G. Saccomando, J. Svensson, and A. Sannino, "Improving voltage disturbance rejection for variable-speed wind turbines," *IEEE Trans. Energy Conversion*, vol. 17, no. 3, pp. 422–427, 2002.

46. E. N. Hinrichren, "Controls for variable pitch wind turbine generators," *IEEE Trans. Power Appar. Syst.*, vol. 103, pp. 886–892, 1984.

47. D. B. Hernan and R. J. Mantz, "Dynamical variable structure controller for power regulation of wind energy conversion systems," *IEEE Trans. Energy Conversion*, vol. 19, no. 4, pp. 756–763, 2004.

48. E. S. Abdin and W. Xu, "Control design and dynamic performance analysis of wind turbine-induction generator unit," *IEEE Trans. Energy Conversion*, vol. 15, no. 1, pp. 91–95, 2000.

49. A. S. Neris, N. A. Vovos, and G. B. Giannakopoulos, "A variable speed wind energy conversion scheme for connection to weak AC systems," *IEEE Trans. Energy Conversion*, vol. 14, no. 1, pp. 122–127, 1999.

50. A. Miller, E. Muljadi, and D. S. Zinger, "A variable speed wind turbine power control," *IEEE Trans. Energy Conversion*, vol. 12, no. 2, pp. 181–186, 1997.

51. E. Gavanidou, A. Bakirtziz, and P. Dokopoulos, "A probabilistic method for the evaluation of the performance of wind diesel energy systems," *IEEE Trans. Energy Conversion*, vol. 6, no. 3, pp. 418–425, 1992.

chapter 3

Electric Machines and Power Systems

3.1 Introduction

Electric machines are necessary for electric power generation. Wind energy is captured by the wind turbine, which acts as the prime mover of the wind generator. Although induction machines are mostly used as wind generators, synchronous machines are also used as wind generators, especially for variable speed operation. Therefore, to understand the wind generator systems, the study of electric machines is necessary. Again, power generated by the wind energy system is connected to the power grid through transformers. Circuit breakers are located on the lines to respond to the fault situations. As a whole, it is essential when studying wind energy systems to have a basic understanding of electric machines and power systems together with the associated controls. This chapter briefly describes the basics of direct current (DC) machines, synchronous machines, induction machines, transformers, circuit breakers, power systems analysis, and power systems control. For detailed study, readers are referred to books dealing with electric machines and power systems [1–4].

3.2 DC Machines

The basic structural features of a DC machine are a stator, which carries the field winding, and a rotor, which carries the armature winding. The stator and rotor together constitute the magnetic circuit or core of the machine, which is a hollow cylinder. The armature is the load-carrying member. The rotor is cylindrical in shape. Armature winding rotates in the magnetic field set up at the stationary winding. An armature winding is continuous; that is, it has no beginning or end. It is composed of a number of coils in series, as shown in Figure 3.1. Depending on the manner in which the coil ends are connected to the commutator bars, armature windings can be grouped into lap windings and wave windings. Wave winding gives greater voltage and smaller current ratings, whereas the lap winding supplies greater current and smaller voltage ratings. Field winding is an exciting system that may be an electrical winding or a permanent magnet and is located on the stator. The coils on the armature

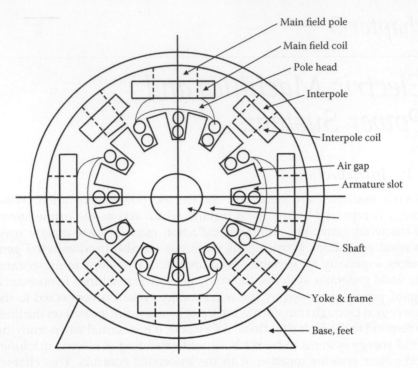

Figure 3.1 Structure of basic DC machine.

are terminated and interconnected through the commutator, which is composed of a number of bars or commutator segments insulated from each other. The commutator rotates with the rotor and serves to rectify the induced voltage and the current in the armature, both of which are alternating currents (AC). Brushes made of conducting carbon graphite are spring loaded to ride on the commutator and act as an interface between the external circuit and the armature winding. The field winding is placed in poles, the number of which is determined by the voltage and current ratings of the machine. For mechanical support, protection from abrasion, and further electrical insulation, nonconducting slot liners are often wedged between the coils and the slot walls. The magnetic material between the slots is called teeth.

In the motoring operation the DC machine is made to work from a DC source and to absorb electrical power. This power is converted into the mechanical form. This is briefly discussed here. If the DC machine's armature that is at rest is connected to a DC source, then a current flows into the armature conductors. If the field is already excited, then these current-carrying conductors experience a force, and the armature experiences a torque. If the restraining torque could be neglected the armature starts rotating in the direction of the force. The conductors now move under the field and cut the magnetic flux, and hence an induced electromotive force (emf) appears in them. The polarity of the induced emf is such as to oppose the cause of the current, which in the present case is the applied voltage. Thus, a back emf appears and tries to reduce the current. As the induced emf and the current act in opposing sense, the machine acts like a sink to the electrical power that the source supplies. This absorbed electrical power gets converted into mechanical form. Thus, the same electrical machine works as a generator of electrical power or the absorber of electrical power depending on the operating condition. The absorbed power gets converted into electrical or mechanical power.

3.3 AC Machines

AC machines produce AC voltage. There are mainly two types of AC machines: (1) synchronous machine; and (2) asynchronous or induction machine.

3.3.1 Synchronous Machines

The synchronous machine is an important electromechanical energy converter. Synchronous generators usually operate together (or in parallel), forming a large power system supplying electrical energy to the loads or consumers. For these applications synchronous machines are built as large units, and their rating ranges from tens to hundreds of megawatts. For high-speed machines, the prime movers are usually steam turbines employing fossil or nuclear energy resources. Low-speed machines are often driven by hydroturbines that employ water power for generation. Smaller synchronous machines are sometimes used for private generation and as standby units, with diesel engines or gas turbines as prime movers.

Synchronous machines can also be used as motors, but they are usually built in very large sizes. The synchronous motor operates at a precise synchronous speed and hence is a constant-speed motor. Unlike the induction motor, whose operation always involves a lagging power factor, the synchronous motor possesses a variable-power-factor characteristic and hence is suitable for power-factor correction applications. A synchronous motor operating without mechanical load is called a compensator. It behaves as a variable capacitor when the field is overexcited and as a variable inductor when the field is underexcited. It is often used in critical positions in a power system for reactive power control.

According to the arrangement of the field and armature windings, synchronous machines may be classified as rotating-armature type or rotating-field type.

Rotating-armature type: The armature winding is on the rotor, and the field system is on the stator. The generated current is brought out to the load via three (or four) slip rings. Insulation problems and the difficulty involved in transmitting large currents via the brushes limit the maximum power output and the generated electromagnetic field. This type is used only in small units, and its main application is as the main exciter in large alternators with brushless excitation systems.

Rotating-field type: The armature winding is on the stator, and the field system is on the rotor. Field current is supplied from the exciter via two slip rings, whereas the armature current is directly supplied to the load. This type is employed universally since very high power can be delivered. Unless otherwise stated, the subsequent discussion refers specifically to rotating-field type synchronous machines.

According to the shape of the field, synchronous machines may be classified as cylindrical-rotor (nonsalient pole) machines and salient-pole machines, shown in Figures 3.2 and 3.3, respectively.

The cylindrical-rotor construction is used in generators that operate at high speeds, such as steam-turbine generators (usually two-pole machines). This type of machine usually has a small diameter-to-length ratio to avoid excessive mechanical stress on the rotor due to the large centrifugal forces.

The salient-pole construction is used in low-speed AC generators (e.g., hydroturbine generators) and also in synchronous motors. This type of machine usually has a large number of poles for low-speed operation and a large diameter-to-length ratio. The field coils are wound on the bodies of projecting poles. A damper winding (which is a partial squirrel-cage

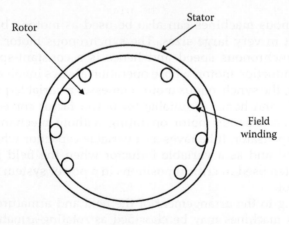

Figure 3.2 Construction of cylindrical-rotor synchronous machine.

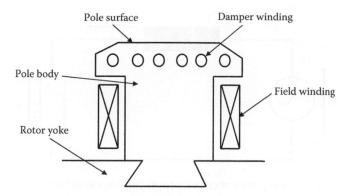

Figure 3.3 Salient-pole rotor construction.

winding) is usually fitted into slots at the pole surface for synchronous motor starting and for improving the stability of the machine.

3.3.1.1 Principle of Cylindrical-Rotor Synchronous Generators

When a synchronous generator is excited with field current and is driven at a constant speed, a balanced voltage is generated in the armature winding. If a balanced load is now connected to the armature winding, a balanced armature current at the same frequency as the emf will flow. Since the frequency of generated emf is related to the rotor speed and the speed of the armature rotating magnetomotive force (mmf) is related to the frequency of the current, it follows that the armature mmf rotates synchronously with the rotor field. An increase in rotor speed results in a rise in the frequency of emf and current, whereas the power factor is determined by the nature of the load.

The effect of the armature mmf on the resultant field distribution is called armature reaction. Since the armature mmf rotates at the same speed as the main field, it produces a corresponding emf in the armature winding. For steady-state performance analysis, the per-phase equivalent circuit shown in Figure 3.4 is used. The effects of armature reaction and armature winding leakage are considered to produce an equivalent internal voltage drop across the synchronous reactance Xs, whereas the field excitation is accounted for by the open-circuit armature voltage E_f. The impedance $Zs = (R + jXs)$ is known as the synchronous impedance of the synchronous generator, where R is the armature resistance.

The circuit equation of the synchronous generator is

$$E_f = V + I.Zs \qquad (3.1)$$

Figure 3.5 shows a voltage phasor diagram of a cylindrical-rotor synchronous generator supplying a lagging-power-factor load. Due to the

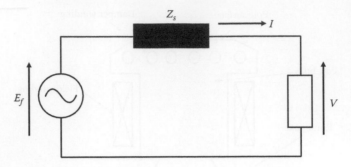

Figure 3.4 Per-phase equivalent circuit of synchronous generator.

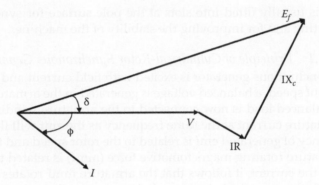

Figure 3.5 Phasor diagram of cylindrical-rotor synchronous generator supplying a lagging-power-factor load (δ, load angle; φ, power factor angle).

synchronous impedance drop, the terminal voltage is less than the open-circuit voltage E_f. For generator operation, the E_f phasor leads the terminal voltage phasor V by the angle δ, often referred to as the load angle.

In practice, synchronous generators seldom operate in the isolated mode. A large number of synchronous machines are usually connected in parallel to supply the loads, forming a large power system known as a *grid*. The voltage and the frequency of the grid remain substantially constant. When a synchronous generator is connected to the grid, its rotor speed and terminal voltage are fixed by the grid and it is said to be operating on infinite busbars. In general, a change in field excitation will result in a change in the operating power factor, whereas a change in mechanical power input will cause a corresponding change in the electrical power output.

The process of paralleling a synchronous machine onto infinite busbars is known as synchronizing. Before a synchronous generator

Table 3.1 Synchronous Generator Parameters

Parameters	Unit
Stator resistance, r_a	Pu
Stator reactance, x_a	Pu
Direct-axis synchronous reactance, X_d	Pu
Quadrature-axis synchronous reactance, X_q	Pu
Direct-axis transient reactance, X'_d	Pu
Quadrature-axis transient reactance, X'_q	Pu
Direct-axis subtransient reactance, X''_d	Pu
Quadrature-axis subtransient reactance, X''_q	Pu
Direct-axis open circuit transient time constant, T'do	sec
Direct-axis open circuit subtransient time constant, T"do	sec
Quadrature-axis open circuit subtransient time constant, T"qo	sec
Inertia constant, H	sec

can be synchronized onto live busbars, the following conditions must be satisfied:

- The voltage of the generator must be equal to that of the busbars.
- The frequency of the generator must be equal to that of the busbars.
- The phase sequence of the generator must be the same as that of the busbars.
- At the instant of synchronizing, the voltage phasors of the generator and the busbars must coincide.

Synchronizing may be achieved with the help of synchronizing lamps, and the rotary lamp method is the most popular. Alternatively, a device known as the synchroscope may conveniently be used to facilitate synchronizing.

Table 3.1 shows the typical synchronous machine parameters when used in wind turbine modeling.

3.3.1.2 Automatic Voltage Regulator System

The ability of a power system to maintain stability depends to a large extent on the controls available on the system to damp the electromechanical oscillations. Hence, the study and design of the automatic voltage regulator (AVR) and governor (GOV) control system models are very important. The AVR controls the excitation voltage of the synchronous machine and attempts to keep constant terminal voltage.

3.3.1.3 Governor Control System

The GOV controls mechanical input power of the synchronous machine and attempts to keep constant angular velocity.

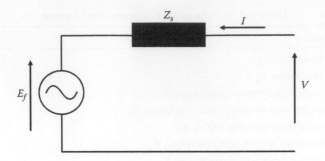

Figure 3.6 Equivalent circuit of cylindrical-rotor synchronous motor.

3.3.1.4 Power System Stabilizer

The function of a power system stabilizer (PSS) is to add damping to the generator rotor oscillations. The PSS uses auxiliary stabilizing signals to control the excitation system to improve power system dynamic performance. Commonly used input signals to the PSS are shaft speed, terminal frequency, and power. Power system dynamic performance is improved by the damping of system oscillations. This is a very effective method of enhancing small-signal stability performance.

3.3.1.5 Operating Principle of Synchronous Motors

A synchronous motor develops a constant torque only when the field system and the armature mmf rotate in synchronism. When the motor is fed from the grid, the supply frequency is constant, and the motor must run at synchronous speed. The synchronous motor is thus a constant-speed motor. The steady-state performance characteristics of the synchronous motor may be studied using the equivalent circuit shown in Figure 3.6. Comparing this with Figure 3.4, it should be noted that the direction of the armature current has been reversed. The circuit equation for a synchronous motor is thus

$$V = E_f + I.Zs \qquad (3.2)$$

To satisfy this circuit equation (3.2), the phasor E_f (often regarded as the back emf of the motor) must lag the terminal voltage V by the load angle δ.

The synchronous motor may be used as a constant-speed drive, particularly for ratings exceeding 15 kW, for example, motor-generator sets for a DC power system, compressors, fans, and blowers. Large synchronous motors have higher efficiencies and can operate at unity power factor; hence, they are smaller in size and weight than induction motors of the same rating. Synchronous motors can also be used for power-factor correction in an industrial plant consisting of a large number of induction motors. An overexcited synchronous motor without mechanical load behaves as a variable capacitor that can be used for reactive power control in a large power system. For the latter application, the machine is often referred to as a synchronous capacitor or a compensator.

3.3.1.6 Permanent Magnet Synchronous Generator

This is a type of synchronous generator where the excitation field is provided by a permanent magnet instead of a coil [1].

Advantages of permanent magnets in synchronous generators are as follows:

- They are more stable and secure during normal operation and do not require an additional DC supply for the excitation circuit.
- They avoid the use of slip rings and hence are simpler and maintenance free.
- They have higher power coefficient and efficiency.
- Synchronous generators are suitable for high capacities, and asynchronous generators, which consume more reactive power, are suitable for smaller capacities.
- Voltage regulation is possible in synchronous generators but not in induction types.
- Condensers are not required for maintaining the power factor in synchronous generators but are necessary for induction generators.
- Because of high coercivity of high-performance permanent magnet materials, such as neodymium, air-gap depth is more tolerable, which puts lower structural constraints on frame and bearing assemblies.

Disadvantages of permanent magnets in synchronous generators are as follows:

- Unlike mmf produced flux density in a winding, the flux density of high-performance permanent magnets, such as derivatives of neodymium and samarium-cobalt, is limited regardless of high coercivity. After all, permanent magnets are magnetized with the higher flux density of an electromagnet. Furthermore, all electric machines are designed to the magnetic core saturation constraints.
- Torque current mmf vectorially combines with the persistent flux of permanent magnets, which leads to higher air-gap flux density and eventually to core saturation.
- Uncontrolled air-gap flux density leads to overvoltage and poor electronic control reliability.
- A persistent magnetic field imposes safety issues during assembly or field service or repair, such as physical injury or electrocution.

- In all cases, high-performance permanent magnet materials are always expensive.
- The mining of high-performance permanent magnet materials is environmentally demanding, and as a result the use of permanent magnets is by no means environmentally friendly.
- Air-gap depth tolerance improves only 20% over other electric machines before magnetic leakage becomes the same concern for any electric machine.
- High-performance permanent magnets have structural and thermal issues.

3.3.1.7 Multimass Shaft System of Synchronous Generator

Usually the turbine-generator is assumed to be made of a single mass. However, in reality, a turbine-generator rotor has a very complex mechanical structure consisting of several predominant masses (e.g., rotors of turbine sections, generator rotor, couplings, and exciter rotor) connected by shafts of finite stiffness. Therefore, when the generator is perturbed, torsional oscillations result between different sections of the turbine-generator rotor. The torsional oscillations in the subsynchronous range could, under certain conditions, interact with the electrical system in an adverse manner. Again, certain electrical system disturbances can significantly reduce the life expectancy of turbine shafts. Therefore, sufficient damping is needed to reduce turbine shaft torsional oscillations. Figure 3.7 shows the turbine-generator shaft model has six masses: a high-pressure (HP) turbine; an intermediate-pressure (IP) turbine; two low-pressure turbines (LPA, LPB); the generator (GEN); and an exciter (EXC) [2].

3.3.2 Asynchronous Machines

An induction motor or asynchronous motor is a type of alternating current motor where power is supplied to the rotor by means of electromagnetic induction. An electric motor turns because of magnetic force exerted between a stationary electromagnet called the stator and a rotating electromagnet called the rotor. Different types of electric motors are distinguished by how electric current is supplied to the moving rotor. In a DC motor and a slip ring AC motor, current is provided to the rotor directly

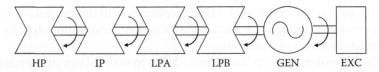

HP   IP   LPA   LPB   GEN   EXC

Figure 3.7 Turbine-generator shaft model.

through sliding electrical contacts called commutators and slip rings. In an induction motor, by contrast, the current is induced in the rotor without contacts by the magnetic field of the stator through electromagnetic induction. An induction motor is sometimes called a rotating transformer because the stator (stationary part) is essentially the primary side of the transformer, and the rotor (rotating part) is the secondary side. Unlike the normal transformer, which changes the current by using time-varying flux, induction motors use rotating magnetic fields to transform the voltage. The current in the primary side creates an electromagnetic field that interacts with the electromagnetic field of the secondary side to produce a resultant torque, thereby transforming the electrical energy into mechanical energy. Induction motors are widely used, especially polyphase induction motors, which are frequently used in industrial drives.

Induction motors are now the preferred choice for industrial motors due to their rugged construction, absence of brushes (which are required in most DC motors), and—thanks to modern power electronics—the ability to control the speed of the motor.

The basic difference between an induction motor and a synchronous AC motor with a permanent magnet rotor is that in the latter the rotating magnetic field of the stator will impose an electromagnetic torque on the magnetic field of the rotor causing it to move (about a shaft), and a steady rotation of the rotor is produced. It is called synchronous because at steady-state the speed of the rotor is the same as the speed of the rotating magnetic field in the stator.

By contrast, the induction motor does not have any permanent magnets on the rotor; instead, a current is induced in the rotor. To achieve this, stator windings are arranged around the rotor so that when energized with a polyphase supply they create a rotating magnetic field pattern that sweeps past the rotor. This changing magnetic field pattern induces current in the rotor conductors. These currents interact with the rotating magnetic field created by the stator and in effect cause a rotational motion on the rotor.

However, for these currents to be induced, the speed of the physical rotor must be less than the speed of the rotating magnetic field in the stator (the synchronous frequency n_s); otherwise, the magnetic field will not be moving relative to the rotor conductors, and no currents will be induced. If by some chance this happens, the rotor typically slows slightly until a current is reinduced, and then the rotor continues as before. This difference between the speed of the rotor and speed of the rotating magnetic field in the stator is called slip. It is unitless and is the ratio between the relative speed of the magnetic field as seen by the rotor (the slip speed) to the speed of the rotating stator field. Due to this, an induction motor is sometimes referred to as an asynchronous machine.

3.3.2.1 Synchronous Speed

It is important to understand the way the behavior of induction motors differs from synchronous motors. The synchronous motors always run at a synchronous speed—a shaft rotation frequency that is an integer fraction of the supply frequency. The synchronous speed of an induction motor is the same fraction of the supply.

It can be shown that the synchronous speed of a motor is determined by

$$n_s = \frac{120 \times f}{p} \tag{3.3}$$

where n_s is the (synchronous) speed of the rotor (in rpm), f is the frequency of the AC supply (in Hz), and p is the number of magnetic poles per phase. (Note: Some texts refer to p as number of pole pairs per phase.)

For example, a six-pole motor operating on 60 Hz power would have a speed of

$$n_s = \frac{120 \times 60}{6} = 1200 \text{ rpm}$$

For example, a six-pole motor, operating on 60 Hz power, would have three pole pairs. The equation of synchronous speed then becomes

$$n_s = \frac{60 \times f}{p} \tag{3.4}$$

where p is the number of pole pairs per phase.

3.3.2.2 Slip

The *slip* is a ratio relative to the synchronous speed and is calculated using

$$s = \left(\frac{n_s - n_r}{n_s} \right) \tag{3.5}$$

where s is the slip, usually between 0 and 1, n_r is the rotor rotation speed (rpm), and n_s is the synchronous rotation speed (rpm).

There are three types of rotor of an induction motor:

Squirrel-cage rotor: The most common rotor, it is made up of bars of either solid copper (most common) or aluminum that span the length of the rotor, and those solid copper or aluminium strips can be shorted or connected by a ring or sometimes not; that is, the rotor can be closed or semiclosed. The rotor bars in squirrel-cage induction motors are not straight but have some skew to reduce noise and harmonics.

Slip ring rotor: A slip ring rotor replaces the bars of the squirrel-cage rotor with windings that are connected to slip rings. When these slip rings are shorted, the rotor behaves similarly to a squirrel-cage rotor; they can also be connected to resistors to produce a high-resistance rotor circuit, which can be beneficial in starting.

Solid-core rotor: A rotor can be made from a solid mild steel. The induced current causes the rotation.

3.3.2.3   Induction Generator or Asynchronous Generator

An induction generator or asynchronous generator is a type of AC electrical generator that uses the principles of induction motors to produce power. Induction generators operate by mechanically turning their rotor in generator mode, giving negative slip. In most cases, a regular AC asynchronous motor is used as a generator, without any internal modifications.

3.3.2.3.1   Principle of Operation.   Induction generators and motors produce electrical power when their shaft is rotated faster than the *synchronous frequency*. For a typical four-pole motor (two pairs of poles on stator) operating on a 60 Hz electrical grid, synchronous speed is 1,800 rotations per minute. A similar four-pole motor operating on a 50 Hz grid will have synchronous speed equal to 1,500 rpm. In normal motor operation, stator flux rotation is faster than the rotor rotation. This is causing stator flux to induce rotor currents, which creates rotor flux with magnetic polarity opposite to the stator. In this way, the rotor is dragged along behind stator flux by value equal to slip.

In generator operation, a certain prime mover (turbine engine) is driving the rotor above the synchronous speed. Stator flux still induces currents in the rotor; however, since the opposing rotor flux is now cutting the stator coils, active current is produced in stator coils, and the motor is now operating as a generator and sending power back to the electrical grid.

Induction generators are not, in general, self-exciting, meaning they require an electrical supply, at least initially, to produce the rotating magnetic flux (although in practice an induction generator will often self-start due to residual magnetism). The electrical supply can be supplied from the electrical grid or, once it starts producing power, from the generator itself. The rotating magnetic flux from the stator induces currents in the rotor, which also produces a magnetic field. If the rotor turns slower than the rate of the rotating flux, the machine acts like an induction motor. If the rotor is turned faster, it acts like a generator, producing power at the synchronous frequency.

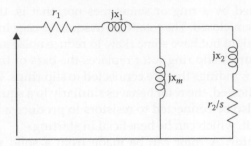

Figure 3.8 Steady-state equivalent circuit of single squirrel-cage induction machine.

Active power delivered to the line is proportional to slip above the synchronous speed. Full-rated power of the generator is reached at very small slip values (motor dependent, typically 3%). At synchronous speed of 1,800 rpm, the generator will produce no power. When the driving speed is increased to 1,860 rpm, full output power is produced. If the prime mover is unable to produce enough power to fully drive the generator, speed will remain somewhere between 1,800 and 1,860 rpm. Figure 3.8 shows a steady-state equivalent circuit of a single squirrel-cage induction machine model, where s denotes a rotational slip.

3.3.2.3.2 Required Capacitance. A capacitor bank must supply reactive power to the motor when used in stand-alone mode. Reactive power supplied should be equal to or greater than the reactive power that machine normally draws when operating as a motor. Terminal voltage will increase with capacitance but is limited by iron saturation.

3.3.2.3.3 Grid and Stand-Alone Connections In induction generators, the magnetizing flux is established by a capacitor bank connected to the machine in stand-alone systems, and in grid connections it draws a magnetizing current from the grid. For a grid-connected system, frequency and voltage at the machine will be dictated by the electric grid, since it is very small compared with the whole system. For stand-alone systems, frequency and voltage are complex functions of machine parameters, capacitance used for excitation, and load value and type.

Induction generators are often used in wind turbines and some micro hydro-installations due to their ability to produce useful power at varying rotor speeds. Induction generators are mechanically and electrically simpler than other generator types. They are also more rugged and require no brushes or commutators. Induction generators are particularly suitable and usually used for wind-generating stations since in this case speed is always a variable factor and the generator is easy on the gearbox. Table 3.2

Table 3.2 Induction Generator Parameters

Parameters	Unit
Stator resistance, r_1	Pu
Stator leakage reactance, x_1	Pu
Magnetizing reactance, X_{mu}	Pu
Rotor resistance, r_2	Pu
Rotor leakage reactance, x_2	Pu
Inertia constant, H	sec

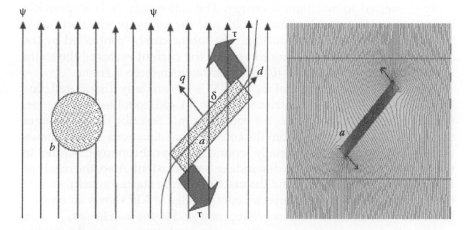

Figure 3.9 An object with anisotropic geometry (a) and isotropic geometry (b) in a magnetic field y and torque production mechanism.

shows the typical induction machine parameters when used in wind turbine modeling.

3.3.3 Synchronous Reluctance Machine

The synchronous reluctance machine (SynRM) uses the reluctance concept and rotating sinusoidal mmf, which can be produced by the traditional induction motor stator, for torque production [5–8]. The reluctance torque concept has a very old history and can be traced back to before 1900. The main idea can be explained by Figure 3.9. In this figure object (a) with an isotropic magnetic material has different (geometric) reluctances on the d-axis and the q-axis, whereas the isotropic magnetic material in object (b) has the same reluctance in all directions. A magnetic field ψ that is applied to the anisotropic object (a) is producing torque if there is an angle difference between the d-axis and the field ($\delta \neq 0$). It is obvious that if the d-axis of object (a) is not aligned with the field, it will introduce a field distortion in the main field. The main direction of this distortion field is aligned along the q-axis of the object. In the SynRM field (ψ) is produced by a sinusoidally distributed winding in a slotted stator, and it links the stator and rotor through a small air gap, exactly as in a traditional IM. The field is rotating at synchronous speed and can be assumed to have a sinusoidal distribution.

In this situation there will always be a torque that acts to reduce the whole system potential energy by reducing the distortion field in the q-axis ($\delta \to 0$). If (δ) load angle is kept constant—for example, by control or applying a load torque—then electromagnetic energy will be continuously converted to mechanical energy. The stator current is responsible for both the magnetization (main field) and the torque production that is trying to reduce the field distortion. The torque can be controlled by the current angle, which is the angle between the current vector of the stator winding and the rotor d-axis (θ) in synchronous reference frame.

Since the stator winding of the SynRM is sinusoidally distributed, flux harmonics in the air gap contribute only to an additional term in the stator leakage inductance. Hence, the equations that describe the behavior of the SynRM can be derived from the conventional equations—Park's equations—depicting a conventionally wound field synchronous machine. In the SynRM, the excitation (field) winding is nonexistent. Also, the machine cage in the rotor is omitted, and the machine can be started synchronously from a standstill by proper inverter control. Figure 3.10 shows an equivalent vector circuit of SynRM including rotor and stator iron losses.

Synchronous reluctance machines use a reluctance rotor design but a standard three-phase AC winding. An induction cage on the rotor provides start torque, but once the machine is close to synchronous speed the reluctance torque will synchronize the machine with the supply. Although synchronous reluctance motors have a low torque capability, they can be used in a number of specific applications such as with clocks and textiles, where precise speed control is required to spin constant-thickness fibers. (This also applies to optical fiber production.)

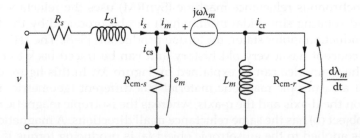

Figure 3.10 Equivalent vector circuit of SynRM including rotor and stator iron losses.

3.3.4   Transformer

A transformer is a static device that transfers electrical energy from one circuit to another through inductively coupled conductors—the transformer's coils. A varying current in the first or primary winding creates a varying magnetic flux in the transformer's core and thus a varying magnetic field through the secondary winding. This varying magnetic field induces a varying emf or "voltage" in the secondary winding. This effect is called mutual induction.

If a load is connected to the secondary, an electric current will flow in the secondary winding and electrical energy will be transferred from the primary circuit through the transformer to the load. In an ideal transformer, the induced voltage in the secondary winding (V_s) is in proportion to the primary voltage (V_p) and is given by the ratio of the number of turns in the secondary (N_s) to the number of turns in the primary (N_p) as follows:

$$\frac{V_s}{V_p} = \frac{N_s}{N_p} \tag{3.6}$$

By appropriate selection of the ratio of turns, a transformer thus allows an AC voltage to be "stepped up" by making N_s greater than N_p or "stepped down" by making N_s less than N_p.

In the vast majority of transformers, the windings are coils wound around a ferromagnetic core, with air-core transformers being a notable exception. Transformers range in size from a thumbnail-sized coupling transformer hidden inside a stage microphone to huge units weighing hundreds of tons used to interconnect portions of power grids. All operate with the same basic principles, although the range of designs is wide. Although new technologies have eliminated the need for transformers in some electronic circuits, transformers are still found in nearly all electronic devices designed for household ("mains") voltage. Transformers are essential for high-voltage electric power transmission, which makes long-distance transmission economically practical.

3.3.4.1   Basic Principles

The transformer is based on two principles: first, that an electric current can produce a magnetic field (electromagnetism); and second, that a changing magnetic field within a coil of wire induces a voltage across the ends of the coil (electromagnetic induction). Changing the current in the primary coil changes the magnetic flux that is developed. The changing magnetic flux induces a voltage in the secondary coil.

An ideal transformer is shown in Figure 3.11. Current passing through the primary coil creates a magnetic field. The primary and secondary coils

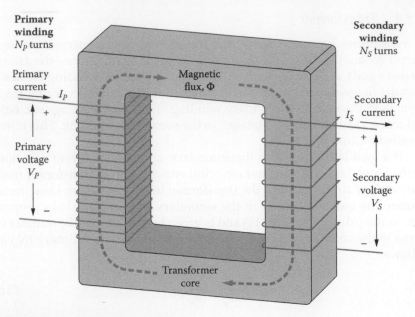

Figure 3.11 An ideal transformer.

are wrapped around a core of very high magnetic permeability, such as iron, so that most of the magnetic flux passes through both the primary and secondary coils.

3.3.4.2 Induction Law

The voltage induced across the secondary coil may be calculated from Faraday's law of induction, which states that

$$V_s = N_s \frac{d\Phi}{dt} \tag{3.7}$$

here V_s is the instantaneous voltage, N_s is the number of turns in the secondary coil, and Φ is the magnetic flux through one turn of the coil. If the turns of the coil are oriented perpendicular to the magnetic field lines, the flux is the product of the magnetic flux density B and the area A through which it cuts. The area is constant, being equal to the cross sectional area of the transformer core, whereas the magnetic field varies with time according to the excitation of the primary. Since the same magnetic flux passes through both the primary and secondary coils in an ideal transformer, the instantaneous voltage across the primary winding equals

$$V_p = N_p \frac{d\Phi}{dt} \tag{3.8}$$

Taking the ratio of the two equations for V_s and V_p gives the basic equation for stepping up or stepping down the voltage:

$$\frac{V_s}{V_p} = \frac{N_s}{N_p} \tag{3.9}$$

N_p/N_s is known as the turns ratio and is the primary functional characteristic of any transformer. In the case of step-up transformers, this may sometimes be stated as the reciprocal, N_s/N_p. Turns ratio is commonly expressed as an irreducible fraction or ratio. For example, a transformer with primary and secondary windings of 100 and 150 turns, respectively, is said to have a turns ratio of 2:3 rather than 0.667 or 100:150.

3.3.4.3   Ideal Power Equation

Figure 3.12 shows an ideal transformer as a circuit element. If the secondary coil is attached to a load that allows current to flow, electrical power is transmitted from the primary circuit to the secondary circuit. Ideally, the transformer is perfectly efficient; all the incoming energy is transformed from the primary circuit to the magnetic field and into the secondary circuit. If this condition is met, the incoming electric power must equal the outgoing power:

$$P_{incoming} = I_p V_p = P_{outgoing} = I_s V_s \tag{3.10}$$

giving the ideal transformer equation

$$\frac{V_s}{V_p} = \frac{N_s}{N_p} = \frac{I_p}{I_s} \tag{3.11}$$

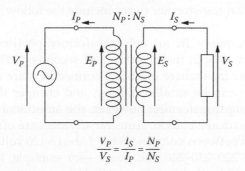

Figure 3.12 The ideal transformer as a circuit element.

Transformers normally have high efficiency, so this formula is a reasonable approximation. If the voltage is increased, then the current is decreased by the same factor. The impedance in one circuit is transformed by the *square* of the turns ratio. For example, if an impedance Z_s is attached across the terminals of the secondary coil, it appears to the primary circuit to have an impedance of $(N_p/N_s)^2 Z_s$. This relationship is reciprocal, so that the impedance Z_p of the primary circuit appears to the secondary to be $(N_s/N_p)^2 Z_p$.

3.3.4.4   Detailed Operation

The previously given simplified description neglects several practical factors, in particular the primary current required to establish a magnetic field in the core and the contribution to the field due to current in the secondary circuit.

Models of an ideal transformer typically assume a core of negligible reluctance with two windings of zero resistance. When a voltage is applied to the primary winding, a small current flows, driving flux around the magnetic circuit of the core. The current required to create the flux is termed the *magnetizing current*; since the ideal core has been assumed to have near-zero reluctance, the magnetizing current is negligible, although it is still required to create the magnetic field.

The changing magnetic field induces an emf across each winding. Since the ideal windings have no impedance, they have no associated voltage drop, so the voltages V_P and V_S measured at the terminals of the transformer are equal to the corresponding emfs. The primary emf, acting as it does in opposition to the primary voltage, is sometimes termed the back emf. This is due to Lenz's law, which states that the induction of emf would always be such that it will oppose development of any such change in magnetic field.

3.3.4.4.1 Types   A wide variety of transformer designs are used for different applications, though they share several common features. Important common transformer types include the following:

1. **Autotransformers:** In an autotransformer portions of the same winding act as both the primary and secondary. The winding has at least three taps where electrical connections are made. An autotransformer can be smaller, lighter, and cheaper than a standard dual-winding transformer; however, the autotransformer does not provide electrical isolation. Autotransformers are often used to step up or down between voltages in the 110–117–120 volt range and voltages in the 220–230–240 volt range—for example, to output either 110 or 120 V (with taps) from 230 V input, allowing equipment from a 100 or 120 V region to be used in a 230 V region. A variable

autotransformer is made by exposing part of the winding coils and making the secondary connection through a sliding brush, giving a variable turns ratio. Such a device is often referred to by the trademark name variac.

2. **Three-phase transformer:** In these types of transformers, a bank of three individual single-phase transformers can be used, or all three phases can be incorporated as a single three-phase transformer. In this case, the magnetic circuits are connected together, and the core thus contains a three-phase flow of flux. A number of winding configurations are possible, giving rise to different attributes and phase shifts. One particular polyphase configuration is the zigzag transformer, used for grounding and in the suppression of harmonic currents.

3. **Leakage transformers:** Also called stray-field transformers, these have a significantly higher leakage inductance than other transformers, sometimes increased by a magnetic bypass or shunt in its core between primary and secondary, which is sometimes adjustable with a set screw. This provides a transformer with an inherent current limitation due to the loose coupling between its primary and the secondary windings. The output and input currents are low enough to prevent thermal overload under all load conditions—even if the secondary is shorted. Leakage transformers are used for arc welding and high-voltage discharge lamps (neon lights and cold cathode fluorescent lamps, which are series-connected up to 7.5 kV AC). They act then both as a voltage transformer and as a magnetic ballast. Other applications are short-circuit-proof extra-low voltage transformers for toys or doorbell installations.

4. **Resonant transformers:** These are a kind of leakage transformer. They use the leakage inductance of their secondary windings in combination with external capacitors to create one or more resonant circuits. Resonant transformers such as the Tesla coil can generate very high voltages and are able to provide much higher current than electrostatic high-voltage generation machines such as the Van de Graaff generator. One of the applications of the resonant transformer is for the Cold Cathode Fluorescent Lamp (CCFL) inverter. Another application of the resonant transformer is to couple between stages of a superheterodyne receiver, where the selectivity of the receiver is provided by tuned transformers in the intermediate frequency amplifiers.

5. **Audio transformers:** These are specifically designed for use in audio circuits. They can be used to block radio frequency interference or the DC component of an audio signal, to split or combine audio signals, or to provide impedance matching between high- and low-impedance circuits, such as between a high-impedance tube (valve)

amplifier output and a low-impedance loudspeaker or between a high-impedance instrument output and the low-impedance input of a mixing console. Such transformers were originally designed to connect different telephone systems to one another while keeping their respective power supplies isolated and are still commonly used to interconnect professional audio systems or system components. Being magnetic devices, audio transformers are susceptible to external magnetic fields such as those generated by AC current-carrying conductors. *Hum* is a term commonly used to describe unwanted signals originating from the "mains" power supply (typically 50 or 60 Hz). Audio transformers used for low-level signals, such as those from microphones, often include shielding to protect against extraneous magnetically coupled signals.

6. **Instrument transformers:** These are used for measuring voltage and current in electrical power systems and for power system protection and control. Where a voltage or current is too large to be conveniently used by an instrument, it can be scaled down to a standardized, low value. Instrument transformers isolate measurement, protection, and control circuitry from the high currents or voltages present on the circuits being measured or controlled.

7. **Current transformers:** These are designed to provide a current in its secondary coil proportional to the current flowing in its primary coil.

8. **Voltage transformers:** (VTs) Also referred to as potential transformers (PTs), these are designed to have an accurately known transformation ratio in both magnitude and phase, over a range of measuring circuit impedances. A voltage transformer is intended to present a negligible load to the supply being measured. The low secondary voltage allows protective relay equipment and measuring instruments to be operated at a lower voltage. Both current and voltage instrument transformers are designed to have predictable characteristics on overloads. Proper operation of overcurrent protective relays requires that current transformers provide a predictable transformation ratio even during a short circuit.

3.3.4.1.2 Classification Transformers can be classified in many different ways, such as the following:

- *By power capacity*: from a fraction of a volt-ampere (VA) to over 1,000 MVA
- *By frequency range*: power, audio, or radio frequency
- *By voltage class*: from a few volts to hundreds of kilovolts
- *By cooling type*: air cooled, oil cooled, fan cooled, or water cooled
- *By application*: such as power supply, impedance matching, output voltage and current stabilizer, or circuit isolation

- *By purpose*: such as distribution, rectifier, arc furnace, or amplifier output
- *By winding turns ratio*: step-up, step-down, isolating with equal or near-equal ratio, variable, multiple windings

3.3.4.4.3   Applications   A major application of transformers is to increase voltage before transmitting electrical energy over long distances through wires. Wires have resistance and so dissipate electrical energy at a rate proportional to the square of the current through the wire. By transforming electrical power to a high-voltage (and therefore low-current) form for transmission and back again afterward, transformers enable economical transmission of power over long distances. Consequently, transformers have shaped the electricity supply industry, permitting generation to be located remotely from points of demand. All but a tiny fraction of the world's electrical power has passed through a series of transformers by the time it reaches the consumer.

Transformers are also used extensively in electronic products to step down the supply voltage to a level suitable for the low-voltage circuits they contain. The transformer also electrically isolates the end user from contact with the supply voltage.

Signal and audio transformers are used to couple stages of amplifiers and to match devices such as microphones and record players to the input of amplifiers. Audio transformers allow telephone circuits to carry on a two-way conversation over a single pair of wires. A balun transformer converts a signal that is referenced to ground to a signal that has balanced voltages to ground, such as between external cables and internal circuits.

3.4   Electrical Power Systems

Electrical power systems mean power networks consisting of generators, cables, transformers, transmission lines, loads, and control systems. Electric power systems vary in size and structural components. However, they all have the same basic characteristics:

- They are composed of three-phase AC systems operating essentially at constant voltage. Generation and transmission facilities use three-phase equipment. Industrial loads are invariably three-phase; single-phase residential and commercial loads are distributed equally among the phases to effectively form a balanced three-phase system.
- They use synchronous machines for generation of electricity. Prime movers convert the primary sources of energy (fossil, nuclear, and hydraulic) to mechanical energy that is, in turn, converted to electrical energy by synchronous generators.

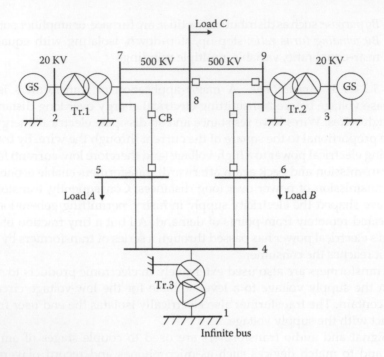

Figure 3.13 Basic elements of a power system.

- They transmit power over significant distances to consumers spread over a wide area. This requires a transmission system comprising subsystems operating at different voltage levels.

Figure 3.13 illustrates the basic elements of a modern power system. Electric power is produced at generating stations (GSs) and transmitted to consumers through a complex network of individual components, including transmission lines, transformers, and switching devices.

This section provides a general overview of conventional power generation systems, then discusses the electric power transmission and distribution systems, introduces power system analysis, and finally describes the power system active and reactive powers control.

3.4.1 Conventional Power Generation

Electricity is produced by generators that convert mechanical energy into electrical energy when large coils are rotated in a powerful magnetic field. Power generation from fossil fuel sources such as petroleum, natural gas, or coal and from large-scale hydroelectric power and nuclear power generation are considered conventional sources. About 80% of commercial power comes from turbine engines powered by steam produced by burning fossil fuels; about 9% is produced using steam from nuclear reactors; about 6% comes from conventional hydroelectric conversion; and about 4% comes from renewable sources (solar, wind, and geothermal). Power plants typically produce between 500 and 900 megawatts of power, or enough to supply the needs of 500,000 to 1 million households. Larger plants require special metals and fabrication and require more downtime for maintenance, and smaller units aren't as economical to operate.

The output of the generation plant is stepped up to a high voltage (typically 155,000 to 765,000 volts) at the transmission substation for connection to the power distribution grid. The grid transports power from multiple power stations to local distribution systems for delivery to homes and businesses.

3.4.2 Electric Power Transmission

Electric power transmission, a process in the delivery of electricity to consumers, is the bulk transfer of electrical power. The transmission system interconnects all major generating stations and main load centers in the system. It forms the backbone of the integrated power system and operates at the highest voltage levels (typically, 230 kV and above). The generator voltages are usually in the range of 11 to 35 kV. These are stepped up to the transmission voltage level, and power is transmitted to transmission substations where the voltages are stepped down to the subtransmission level (typically, 69 kV to 138 kV). The generation and subtransmission subsystems are often referred to as the bulk power system. Capacitor banks and reactor banks are usually installed in the substations for maintaining the transmission line voltage. The elementary diagram of a transmission system is shown in Figure 3.13.

Electric power is transferred from generating stations to consumers through overhead lines and cables. Overhead lines are used for long distances in open country and rural areas, whereas cables are used for underground transmission in urban areas and for underwater crossings. For the same rating, cables are 10 to 15 times more expensive than overhead lines and are used only in special situations where overhead lines cannot be used; the distance in such applications is short.

Transmission lines vary from a few kilometers long in an urban environment to over 1,000 km for lines carrying power from remote hydroelectric plants. They may differ greatly in the amount of power carried. Because requirements vary, many technical, economic, and environmental factors must be considered when new lines are planned.

The basic modes of transmission are DC and AC. In direct current, the current flows in one direction only; in alternating current it reverses its direction many times per second. It is difficult to transform direct current from one voltage to another; hence, initially DC had to be transmitted at the low voltage at which it was generated and used. This fact limited its applicability: If transmission of large amounts of electricity or transmission over long distances was required, the cost of the conductor (copper wire) was prohibitive. Alternating current may be generated at a low voltage, boosted to a higher voltage by a transformer, transmitted, and converted back to a lower voltage before use. Consequently, following the development of the transformer in the 1890s, most electricity was transmitted as AC.

However, DC transmission has a number of advantages and is being more widely used. For example, a DC line, requiring only two conductors instead of the three needed for an AC line, costs about two-thirds as much. Further, in DC transmission the effective voltage is equal to the peak voltage, whereas in AC transmission the peak voltage is 40% higher. Since radio interference increases with the peak voltage and decreases as the conductor size is increased, the DC system can carry a higher effective voltage than an AC line of equivalent size and still maintain an acceptable radio interference level. Thus, in some long lines carrying bulk power from remote generating sites, power is generated as AC, boosted to a high voltage, converted to DC for transmission, then reconverted to AC and transformed to a lower voltage for use. The cost of the converter stations at either end is offset by the lower cost of the line. DC transmission is also advantageous for transmitting power through submarine cables.

Overhead power lines have three major components: support structure, insulation, and conductors. Support structures can be wooden poles, free-standing steel towers, or guyed towers of steel or aluminum. Glass or porcelain suspension insulators have traditionally separated the live conductors from the grounded towers. Each insulator consists of a metal cap on top and a metal pin underneath separated by the glass or porcelain insulation. These units are used to form insulator strings that vary in length depending on the voltage level and application. Several strings may be used in parallel to carry the weight of the conductors. For 735 kV, about 30 insulators are used. New types of insulators have been developed using polymers; field testing and full-scale use became more prevalent during the 1980s. In the early days of electrical transmission, copper was used extensively as a conductor, but now virtually all conductors are aluminum. Each conductor is made of many strands (1–5 mm in diameter) combined to give an overall diameter of 4–50 mm. In most conductors, steel or a high-strength aluminum alloy is used for the core strands to give the conductor added strength. In a transmission line, up to four conductors may be used in parallel to form a conductor bundle.

Table 3.3 Typical Overhead Transmission Line Parameters

Nominal Voltage	230 kV	345 kV	500 kV	765 kV	1,100 kV
R (Ω/km)	0.050	0.037	0.028	0.012	0.005
$x_L = \omega L$ (Ω/km)	0.488	0.367	0.325	0.329	0.292
$b_C = \omega C$ (μs/km)	3.371	4.518	5.200	4.978	5.544
α (nepers/km)	0.000067	0.000066	0.000057	0.000025	0.000012
β (rad/km)	0.00128	0.00129	0.00130	0.00128	0.00127
Z_C (Ω)	380	285	250	257	230
SIL (MW)	140	420	1000	2280	5260
Charging MVA/km $=V_0^2 b_C$	0.18	0.54	1.30	2.92	6.71

Note: SIL and charging MVA are three-phase values.
[a] Rated frequency is assumed to be 60 Hz.
[b] Bundled conductors used for all lines listed except for the 230 kV line.
[c] R, x_L, and b_C are per-phase values.

3.4.2.1 Line Parameters for Overhead Transmission Line

A transmission line is characterized by four parameters: (1) series resistance R due to the conductor resistivity; (2) shunt conductance G due to leakage currents between the phases and ground; (3) series inductance L due to the magnetic field surrounding the conductors; and (4) shunt capacitance C due to the electric field between conductors. Table 3.3 gives typical parameters of overhead lines of nominal voltage ranging from 230 kV to 1,100 kV [2].

3.4.3 Electric Power Distribution

The electric power distribution system represents the final stage in the transfer of power to the individual customers. It comprises those parts of an electric power system between the subtransmission system and the consumers' service switches. It includes distribution substations; primary distribution feeders; distribution transformers; secondary circuits, including the services to the consumer; and appropriate protective and control devices. Sometimes, the subtransmission system is also included in the definition. The elementary diagram of a distribution system is shown in Figure 3.13.

The subtransmission circuits of a typical distribution system deliver electric power from bulk power sources to the distribution substations. The subtransmission voltage is usually between 34.5 and 138 kV. The distribution substation, which is made up of power transformers together with the necessary voltage-regulating apparatus, busbars, and switchgear, reduces the subtransmission voltage to a lower primary system voltage for local distribution. The three-phase primary feeder, which usually operates at voltages from 4.16 to 34.5 kV, distributes electric power from the low-voltage bus of the substation to its load center, where it branches into three-phase subfeeders and three-phase and occasionally single-phase laterals. Small industrial customers are supplied by primary feeders at this voltage level. The secondary distribution feeders supply residential and commercial customers at 120/240 V.

Most of the three-phase distribution system lines consist of three-phase conductors and a common or neutral conductor, making a total of four wires. Single-phase branches (made up of two wires) supplied from the three-phase mains provide power to residences, small stores, and farms. Loads are connected in parallel to common power-supply circuits.

3.4.4   Power System Analysis

Modern power systems have grown larger and more geographically expansive with many interconnections between neighboring systems. Proper planning, operation, and control of such large systems require advanced computer-based techniques. Power system analysis is an important tool for evaluating the future development and needed investments in the electricity sector. It can assist power companies, national authorities, and decision makers prepare analyses of large and complex power systems. The power sector analyses can be used to evaluate the consequences of making psychical changes in power systems. This may, for instance, have to do with the closure of existing power plants, the establishment of new power plants, or increased transmission capacity between regions or countries. Power system analysis provides an active knowledge of various numerical techniques that can be applied to the solution of large interconnected power systems.

Power system analysis includes the study of the following:

1. The development of power systems and major components in the power system
2. Three-phase power systems
3. The steady-state presentation and modeling of synchronous machines and transformers
4. Transmission line parameters and their calculations
5. Transmission line modeling and the performance and compensation of transmission lines
6. Network models based on the admittance and impedance representations
7. The power flow problem of a system during normal operation
8. Economic dispatch and the basics of unit commitment

9. Symmetrical and unsymmetrical components
10. Balanced and unbalanced faults analysis
11. Power system state estimation
12. Contingency analysis
13. Power system controls
14. Power system stability problem

3.4.5   Power Flow Study

In power engineering, the power flow study (also known as load flow study) is an important tool involving numerical analysis applied to a power system. Unlike traditional circuit analysis, a power flow study usually uses simplified notation such as a one-line diagram and per-unit system and focuses on various forms of AC power (i.e., reactive, real, and apparent) rather than voltage and current. It analyzes the power systems in normal steady-state operation. There exist a number of software implementations of power flow studies.

In addition to a power flow study, sometimes called the *base case*, many software implementations perform other types of analyses, such as short-circuit fault analysis and economic analysis. In particular, some programs use linear programming to find the optimal power flow, the conditions that give the lowest cost per kilowatt-hour delivered.

Power flow or load flow studies are important for planning future expansion of power systems as well as for determining the best operation of existing systems. The principal information obtained from the power flow study is the magnitude and phase angle of the voltage at each bus and the real and reactive power flowing in each line. Commercial power systems are usually too large to allow for hand solution of the power flow.

The goal of a power flow study is to obtain complete voltage angle and magnitude information for each bus in a power system for specified load and generator real power and voltage conditions. Once this information is known, real and reactive power flow on each branch as well as generator reactive power output can be analytically determined. Due to the nonlinear nature of this problem, numerical methods are employed to obtain a solution that is within an acceptable tolerance.

The solution to the power flow problem begins with identifying the known and unknown variables in the system. The known and unknown variables are dependent on the type of bus. A bus without any generators connected to it is called a *load bus*. With one exception, a bus with at least one generator connected to it is called a *generator bus*. The exception is one arbitrarily selected bus that has a generator. This bus is referred to as the *slack bus*.

3.4.6   Per-Unit System and Base Quantities

In power system computations, great simplifications can be realized by employing a system in which the electrical quantities are expressed as per units of properly chosen base quantities. The per-unit value of any quantity is defined as the ratio of the actual value to its base value [4].

In an electrical circuit, voltage, current, volt-ampere, and impedance are so related that selection of base values of any two of them determines the base values for the remaining two. Usually, base kVA (or MVA) and base voltage in kV are the quantities selected to specify the base. For single-phase systems—or three-phase systems where the term *current* refers to line current, the term *voltage* refers to voltage to neutral, and the term *kVA* refers to kVA per phase—the following formulas relate the various quantities:

$$\text{Base current in amperes} = \text{base kVA/base voltage in kV} \tag{3.12}$$

Base impedance in ohms = base voltage in volts/base current in amperes

$$= (\text{base voltage in kV})^2 \times 1000/\text{base KVA}$$

$$= (\text{base voltage in kV})^2/\text{base MVA} \tag{3.13}$$

Base power in kW = base kVA

Base power in MW = base MVA

$$\text{Per unit quantity} = \text{actual quantity/base quantity} \tag{3.14}$$

Occasionally a quantity may be given in percent, which is obtained by multiplying the per-unit quantity by 100. Base impedance and base current can be computed directly from three-phase values of base kV and base kVA. If we interpret base kVA and base voltage in kV to mean base kVA for the total of the three phases and base voltage from line to line, then

$$\text{Base current in amperes} = \text{base kVA/}\{\sqrt{3} \times \text{base voltage in kV}\} \tag{3.15}$$

Base impedance in ohms = $\{(\text{base voltage in kV}/\sqrt{3})2 \times 1000\}/\text{base kVA/3}$

$$= \{(\text{base voltage in kV})^2 \times 1000\}/\text{base kVA}$$

$$= (\text{base voltage in kV})^2/\text{base MVA} \tag{3.16}$$

Therefore, the same equation for base impedance is valid for either single-phase or three-phase circuits. In the three-phase case line-to-line kV must be used in the equation with three-phase kVA or MVA. Line-to-neutral kV must be used with kVA or MVA per phase.

3.4.6.1   Change of Bases

Per-unit impedance of a circuit element = {(actual impedance in ohms)
        × (base kVA)}/{(base voltage in kV)^2 × 1000}   (3.17)

which shows that per unit impedance is directly proportional to the base kVA and inversely proportional to the square of the base voltage. Therefore, to change from per unit impedance on a given base to per unit impedance on a new base, the following equation applies:

$$\text{Per unit } Z_{\text{new}} = \text{Per unit } Z_{\text{given}} \left(\frac{\text{base kV}_{\text{given}}}{\text{base kV}_{\text{new}}} \right)^2 \left(\frac{\text{base kVA}_{\text{new}}}{\text{base kVA}_{\text{given}}} \right) \quad (3.18)$$

When the resistance and reactance of a device are given by the manufacturer in percent or per unit, the base is understood to be the rated kVA and kV of the apparatus. The ohmic values of resistance and leakage reactance of a transformer depend on whether they are measured on the high- or low-voltage side of the transformer. If they are expressed in per unit, the base kVA is understood to be the kVA rating of the transformer. The base voltage is understood to be the voltage rating of the side of the transformer where the impedance is measured.

$$Z_{LT} = \left(\frac{kV_L}{kV_H} \right)^2 \times Z_{HT} \quad (3.19)$$

where Z_{LT} and Z_{HT} are the impedances referred to the low-voltage and high-voltage sides of the transformer, respectively, and kV_L and kV_H are the rated low-voltage and high-voltage of the transformer, respectively.

$$\therefore Z_{LT} \text{ in per unit} = \frac{(kV_L / kV_H)^2 \times Z_{HT} \times kVA}{(kV_L)^2 \times 1000}$$

$$= \frac{Z_{HT} \times kVA}{(kV_H)^2 \times 1000}$$

$$= Z_{LT} \text{ in per unit}$$

A great advantage in making per-unit computations is realized by the proper selection of different voltage bases for circuits connected to each other through a transformer. To achieve the advantage, the voltage bases for the circuits connected through the transformer must have the same ratio as the turns ratio of the transformer windings.

3.4.6.2 Per-Unit and Percent Admittance

$$Z_{ohm} = \frac{1}{Y_{mho}} \quad \text{and} \quad Y = \frac{1}{Z}$$

$$\text{Base admittance } Y_b = \frac{1}{Z_b} = \frac{MVA_b}{KV_b^2} \tag{3.20}$$

$$\therefore Y_{pu} = \frac{Y}{Y_b} = Y\frac{KV_b^2}{MVA_b} = YZ_b = \frac{Z_b}{Z} = \frac{1}{Z_{pu}} \tag{3.21}$$

$$Z_{percent} = Z_{pu} \times 100, \quad Y_{percent} = Y_{pu} \times 100$$

$$\therefore Y_{percent} = \frac{1}{Z_{pu}} \times 100 = \frac{10^4}{Z_{percent}} \tag{3.22}$$

$$P = \sqrt{3}\,VI\cos\varphi, \quad Q = \sqrt{3}\,VI\sin\varphi$$

$$\therefore P_{pu} = \frac{\sqrt{3}\,VI\cos\varphi}{\sqrt{3}\,V_bI_b} = V_{pu}I_{pu}\cos\varphi$$

$$Q_{pu} = \frac{\sqrt{3}\,VI\sin\varphi}{\sqrt{3}\,V_bI_b} = V_{pu}I_{pu}\sin\varphi \tag{3.23}$$

3.4.7 Faults in Power Systems

In an electric power system, a fault is any abnormal flow of electric current. For example, a short circuit is a fault in which current flow bypasses the normal load. An open-circuit fault occurs if a circuit is interrupted by some failure. In three-phase systems, a fault may involve one or more phases and ground or may occur only between phases. In a *ground fault* or *earth fault*, current flows into the earth. The prospective short-circuit current of a fault can be calculated for power systems. In power systems, protective devices detect fault conditions and operate circuit breakers and other devices to limit the loss of service due to a failure [9–10].

In a polyphase system, a fault that affects all phases equally is known as a *symmetrical fault*. If only some phases are affected, the *asymmetrical fault* requires use of methods such as symmetrical components for analysis, since the simplifying assumption of equal current magnitude in all phases is no longer applicable.

3.4.7.1 Transient Fault

A transient fault is a fault that is no longer present if power is disconnected for a short time. Many faults in overhead power lines are transient in nature. At the occurrence of a fault, power system protection operates to isolate the area of the fault. A transient fault will then clear, and the power line can be returned to service. Typical examples of transient faults include the following:

- Momentary tree contact
- Bird or other animal contact
- Lightning strike
- Conductor clash

In electricity transmission and distribution systems, an automatic reclose function is commonly used on overhead lines to attempt to restore power in the event of a transient fault. This functionality is not as common on underground systems because faults there are typically of a persistent nature. Transient faults may still cause damage both at the site of the original fault or elsewhere in the network as a fault current is generated.

3.4.7.2 Persistent Fault

A persistent fault does not disappear when power is disconnected. Faults in underground power cables are often persistent. Underground power lines are not affected by trees or lightning, so faults, when they occur, are probably due to damage. In such cases, if the line is reconnected it is likely to be damaged only further.

3.4.7.3 Symmetrical Fault

A symmetrical or balanced fault affects each of the three phases equally. In transmission line faults, roughly 5% are symmetrical. This is in contrast to an asymmetrical fault, where the three phases are not affected equally. In practice, most faults in power systems are unbalanced. With this in mind, symmetrical faults can be viewed as somewhat of an abstraction; however, since asymmetrical faults are difficult to analyze, their analysis is built up from a thorough understanding of symmetrical faults.

3.4.7.4   *Asymmetrical Fault*

An asymmetrical or unbalanced fault does not affect each of the three phases equally. Common types of asymmetric faults and their causes are as follows:

- Line-to-line: A short circuit between lines, caused by ionization of air or when lines come into physical contact, for example, due to a broken insulator
- Line-to-ground: A short circuit between one line and ground, very often caused by physical contact, for example, due to lightning or other storm damage
- Double line-to-ground: Two lines come into contact with the ground (and each other), also commonly due to storm damage

3.4.7.5   *Analysis*

Symmetrical faults can be analyzed via the same methods as any other phenomena in power systems, and in fact many software tools exist to accomplish this type of analysis automatically (see Section 3.4.5). However, another method is as accurate and is usually more instructive.

First, some simplifying assumptions are made. It is assumed that all electrical generators in the system are in phase and are operating at the nominal voltage of the system. Electric motors can also be considered to be generators, because when a fault occurs they usually supply rather than draw power. The voltages and currents are then calculated for this base case.

Next, the location of the fault is considered to be supplied with a negative voltage source, equal to the voltage at that location in the base case, while all other sources are set to zero. This method makes use of the principle of superposition.

To obtain a more accurate result, these calculations should be performed separately for three separate time ranges:

- Subtransient is first and is associated with the largest currents.
- Transient comes between subtransient and steady-state.
- Steady-state occurs after all the transients have had time to settle.

An asymmetrical fault breaks the underlying assumptions used in three-phase power, namely, that the load is balanced on all three phases. Consequently, it is impossible to directly use tools such as the one-line diagram, where only one phase is considered. However, due to the linearity of power systems, it is commonplace to consider the resulting voltages and currents as a superposition of symmetrical components, to which three-phase analysis can be applied.

In the method of symmetrical components, the power system is seen as a superposition of three components:

- A positive-sequence component, in which the phases are in the same order as the original system: a–b–c
- A negative-sequence component, in which the phases are in the opposite order as the original system: a–c–b
- A zero-sequence component, which is not truly a three-phase system but instead all three phases are in phase with each other.

To determine the currents resulting from an asymmetrical fault, one must first know the per-unit zero-, positive-, and negative-sequence impedances of the transmission lines, generators, and transformers involved. Three separate circuits are then constructed using these impedances. The individual circuits are next connected together in a particular arrangement that depends on the type of fault being studied (this can be found in most power systems textbooks) [1–5]. Once the sequence circuits are properly connected, the network can then be analyzed using classical circuit analysis techniques. The solution results in voltages and currents that exist as symmetrical components; these must be transformed back into phase values by using the A matrix.

Analysis of the prospective short-circuit current is required for selection of protective devices such as fuses and circuit breakers. If a circuit is to be properly protected, the fault current must be high enough to operate the protective device within as short a time as possible; also, the protective device must be able to withstand the fault current and extinguish any resulting arcs without itself being destroyed or sustaining the arc for any significant length of time.

The magnitudes of fault currents differ widely depending on the type of earthing system used, the installation's supply type and earthing system, and its proximity to the supply. For example, for a 230 V, 60 A, or 120 V/240 V supply, fault currents may be a few thousand amperes. Large low-voltage networks with multiple sources may have fault levels of 300,000 amperes. A high-resistance-grounded system may restrict line-to-ground fault current to only 5 amperes. Prior to selecting protective devices, prospective fault current must be measured reliably at the origin of the installation and at the farthest point of each circuit, and this information could be applied properly to the application of the circuits.

3.4.7.6   Detecting and Locating Faults

Locating faults in a cable system can be done either with the circuit deenergized or in some cases with the circuit under power. Fault location techniques can be broadly divided into terminal methods, which use voltages and currents measured at the ends of the cable, and tracer methods, which require inspection along the length of the cable. Terminal methods can be used to locate the general area of the fault to expedite tracing on a long or buried cable.

In very simple wiring systems, the fault location is often found through visual inspection of the wires. In complex wiring systems (e.g., aircraft wiring) where the electrical wires may be hidden behind cabinets and extended for miles, wiring faults are located with a time-domain reflectometer, which sends a pulse down the wire and then analyzes the returning reflected pulse to identify faults within the electrical wire.

In historic submarine telegraph cables, sensitive galvanometers were used to measure fault currents; by testing at both ends of a faulted cable, the fault location could be isolated to within a few miles, which allowed the cable to be grappled up and repaired. The Murray loop and the Varley loop were two types of connections for locating faults in cables.

Sometimes an insulation fault in a power cable will not show up at lower voltages. A "thumper" test set applies a high-energy, high-voltage pulse to the cable. Fault location is done by listening for the sound of the discharge at the fault. Although this test contributes to damage at the cable site, it is practical because the faulted location would have to be reinsulated when found in any case.

In a high-resistance grounded distribution system, a feeder may develop a fault-to-ground but the system continues in operation. The faulted, but energized, feeder can be found with a ring-type current transformer collecting all the phase wires of the circuit; only the circuit containing a fault-to-ground will show a net unbalanced current. To make the ground fault current easier to detect, the grounding resistor of the system may be switched between two values so that the fault current pulses.

3.4.8 Power System Stability

Successful operation of a power system depends largely on the engineer's ability to provide reliable and uninterrupted service to the consumers. In other words, the power system operator must maintain a very high standard of continuous electrical service. Power system stability may be defined as the property of the system that enables the synchronous machines of the system to respond to a disturbance from a normal operating condition to return to a condition where their operation is again normal. Instability in a power system may be manifested in many different ways depending on the system configuration and operating mode. Traditionally, the stability problem has been one of maintaining synchronous operation. Since power systems rely on synchronous machines for generation of electrical power, a necessary condition for satisfactory system operation is that all synchronous machines remain in synchronism or, colloquially, "in step." This aspect of stability is influenced by the dynamics of generator rotor angles and power–angle relationships. This chapter presents an overview of the power system stability phenomena including physical concepts, classification, and definition of related terms.

3.4.8.1   Classification of Stability

Stability studies are usually classified into three types depending on the nature and order of magnitude of the disturbance. These are rotor angle stability, frequency stability, and voltage stability [9].

3.4.8.1.1   Rotor Angle Stability   Rotor angle stability refers to the ability of synchronous machines of an interconnected power system to remain in synchronism after being subjected to a disturbance. It depends on the ability to maintain and restore equilibrium between electromagnetic torque and mechanical torque of each synchronous machine in the system. Instability that may result occurs in the form of increasing angular swings of some generators leading to their loss of synchronism with other generators.

The rotor angle stability problem involves the study of the electromechanical oscillations inherent in power systems. A fundamental factor in this problem is the manner in which the power outputs of synchronous machines vary as their rotor angles change. Under steady-state conditions, there is equilibrium between the input mechanical torque and the output electromagnetic torque of each generator, and the speed remains constant. If the system is perturbed, this equilibrium is upset, resulting in acceleration or deceleration of the rotors of the machines according to the laws of motion of a rotating body. If one generator temporarily runs faster than another, the angular position of its rotor relative to that of the slower machine will advance. The resulting angular difference transfers part of the load from the slow machine to the fast machine, depending on the power–angle relationship. This tends to reduce the speed difference and hence the angular separation. The power–angle relationship is highly nonlinear. Beyond a certain limit, an increase in angular separation is accompanied by a decrease in power transfer such that the angular separation is increased further. Instability results if the system cannot absorb the kinetic energy corresponding to these rotor speed differences. For any given situation, the stability of the system depends on whether the deviations in angular positions of the rotors result in sufficient restoring torques. Loss of synchronism can occur between one machine and the rest of the system, or between groups of machines, with synchronism maintained within each group after separating from each other. The change in electromagnetic torque of a synchronous machine following a perturbation can be resolved into two components:

- *Synchronizing torque component,* in phase with rotor angle deviation.
- *Damping torque component,* in phase with the speed deviation.

System stability depends on the existence of both components of torque for each of the synchronous machines. Lack of sufficient synchronizing torque results in *aperiodic* or *nonoscillatory instability*, whereas lack of damping torque results in *oscillatory instability*.

For convenience in analysis and for gaining useful insight into the nature of stability problems, it is useful to characterize rotor angle stability in terms of the following two subcategories:

1. *Small-disturbance (or small-signal) rotor angle stability* is concerned with the ability of the power system to maintain synchronism under small disturbances. The disturbances are considered to be sufficiently small that linearization of system equations is permissible for purposes of analysis.
   a. Small-disturbance stability depends on the initial operating state of the system. Instability that may result can be of two forms:
      i. Increase in rotor angle through a nonoscillatory or aperiodic mode due to lack of synchronizing torque
      ii. Rotor oscillations of increasing amplitude due to lack of sufficient damping torque.
   b. In today's power systems, small-disturbance rotor angle stability problems are usually associated with insufficient damping of oscillations. The aperiodic instability problem has been largely eliminated by use of continuously acting generator voltage regulators; however, this problem can still occur when generators operate with constant excitation when subjected to the actions of excitation limiters (field current limiters).
   c. Small-disturbance rotor angle stability problems may be either local or global in nature. Local problems involve a small part of the power system and are usually associated with rotor angle oscillations of a single power plant against the rest of the power system. Such oscillations are called local plant mode oscillations. Stability (damping) of these oscillations depends on the strength of the transmission system as seen by the power plant, generator excitation control systems, and plant output.
   d. Global problems are caused by interactions among large groups of generators and have widespread effects. They involve oscillations of a group of generators in one area swinging against a group of generators in another area. Such oscillations are called interarea-mode oscillations. Their characteristics are very complex and significantly differ from those of local plant mode

oscillations. Load characteristics, in particular, have a major effect on the stability of interarea modes.

  e. The time frame of interest in small-disturbance stability studies is on the order of 10 to 20 seconds following a disturbance.

2. *Large-disturbance rotor angle stability* or *transient stability*, as it is commonly referred to, is concerned with the ability of the power system to maintain synchronism when subjected to a severe disturbance, such as a short circuit on a transmission line. The resulting system response involves large excursions of generator rotor angles and is influenced by the nonlinear power–angle relationship.

  a. Transient stability depends on both the initial operating state of the system and the severity of the disturbance. Instability is usually in the form of aperiodic angular separation due to insufficient synchronizing torque, manifesting as *first swing instability*. However, in large power systems, transient instability may not always occur as first swing instability associated with a single mode; it could be a result of superposition of a slow interarea swing mode and a local-plant swing mode causing a large excursion of rotor angle beyond the first swing. It could also be a result of nonlinear effects affecting a single mode causing instability beyond the first swing.

  b. The time frame of interest in transient stability studies is usually 3 to 5 seconds following the disturbance. It may extend to 10 to 20 seconds for very large systems with dominant interarea swings. Small-disturbance rotor angle stability and transient stability are categorized as *short-term* phenomena.

  c. Transient stability limit: Refers to the maximum flow of power possible through a point in the system without the loss of stability when a sudden disturbance occurs.

  d. Critical clearing time: Related to transient stability; the maximum time between the fault initiation and its clearing such that the power system is transiently stable. This includes relay and breaker operating times and possibly the time elapsed for the trip signal to reach the other end breaker. Clearing times are in the range of a few power-frequency cycles in modern power systems employing high-speed circuit breakers (one-cycle breakers are in service) and solid-state relays.

 e. Dynamic and steady-state stability studies are less extensive in scope and involve one or just a few machines undergoing slow or gradual changes in operating conditions. Therefore, dynamic and steady-state stability studies concern the stability of the locus of essentially steady-state operating points of the system. The distinction made between steady-state and dynamic stability studies is really artificial since the stability problems are the same in nature; they differ only in the degree of detail used to model the machines. In dynamic stability studies, the excitation system and turbine-governing system are represented, along with synchronous machine models, which provide for flux-linkage variation in the machine air-gap. Steady-state stability problems use a very simple generator model that treats the generator as a constant voltage source.

3.4.8.1.2   Frequency Stability   Frequency stability refers to the ability of a power system to maintain steady frequency following a severe system upset resulting in a significant imbalance between generation and load. It depends on the ability to maintain and restore equilibrium between system generation and load with minimum unintentional loss of load. Instability that may result occurs in the form of sustained frequency swings leading to tripping of generating units or loads.

Severe system upsets generally result in large excursions of frequency, power flows, voltage, and other system variables, thereby invoking the actions of processes, controls, and protections that are not modeled in conventional transient stability or voltage stability studies. These processes may be very slow, such as boiler dynamics, or triggered only for extreme system conditions, such as volts/hertz protection tripping generators.

In large interconnected power systems, this type of situation is most commonly associated with conditions following splitting of systems into islands. Stability in this case is a question of whether each island will reach a state of operating equilibrium with minimal unintentional loss of load. It is determined by the overall response of the island as evidenced by its mean frequency rather than relative motion of machines. Generally, frequency stability problems are associated with inadequacies in equipment responses, poor coordination of control and protection equipment, or insufficient generation reserve. Examples of such problems are reported in references. In isolated island systems, frequency stability could be of concern for any disturbance causing a relatively significant loss of load or generation.

During frequency excursions, the characteristic times of the processes and devices that are activated will range from fraction of seconds, corresponding to the response of devices such as underfrequency load shedding and generator controls and protections, to several minutes, corresponding to the response of devices such as prime mover energy supply systems and load voltage regulators. Therefore, frequency stability may be considered a *short-term* phenomenon or a *long-term* phenomenon.

An example of short-term frequency instability is the formation of an undergenerated island with insufficient underfrequency load shedding such that frequency decays rapidly causing blackout of the island within a few seconds. On the other hand, more complex situations in which frequency instability is caused by steam turbine overspeed controls or boiler/reactor protection and controls are longer-term phenomena with the time frame of interest ranging from tens of seconds to several minutes.

During frequency excursions, voltage magnitudes may change significantly, especially for islanding conditions with underfrequency load shedding that unloads the system. Voltage magnitude changes, which may be higher in percentage than frequency changes, affect the load–generation imbalance.

High voltage may cause undesirable generator tripping by poorly designed or coordinated loss of excitation relays or volts/hertz relays. In an overloaded system, low voltage may cause undesirable operation of impedance relays.

3.4.8.1.3 Voltage Stability   Voltage stability refers to the ability of a power system to maintain steady voltages at all buses in the system after being subjected to a disturbance from a given initial operating condition. It depends on the ability to maintain and restore equilibrium between load demand and load supply from the power system. Instability that may result occurs in the form of a progressive fall or rise of voltages of some buses. A possible outcome of voltage instability is loss of load in an area or tripping of transmission lines and other elements by their protective systems leading to cascading outages. Loss of synchronism of some generators may result from these outages or from operating conditions that violate field current limit. Progressive drop in bus voltages can also be associated with rotor angle instability. For example, the loss of synchronism of machines as rotor angles between two groups of machines approach 180 causes rapid drop in voltages at intermediate points in the network close to the electrical center. Normally, protective systems operate to separate the two groups of machines and the voltages recover to levels depending on the postseparation conditions. If, however, the system is not so separated, the voltages near the electrical center rapidly oscillate between high and low values as a result of repeated *pole slips* between the two groups of machines. In contrast, the type of sustained fall of voltage that is related to voltage instability involves loads and may occur where rotor angle stability is not an issue.

The term *voltage collapse* is also often used. It is the process by which the sequence of events accompanying voltage instability leads to a blackout or abnormally low voltages in a significant part of the power system. Stable (steady) operation at low voltage may continue after transformer tap changers reach their boost limit, with intentional or unintentional tripping of some load. Remaining load tends to be voltage sensitive, and the connected demand at normal voltage is not met.

The driving force for voltage instability is usually the loads; in response to a disturbance, power consumed by the loads tends to be restored by the action of motor slip adjustment, distribution voltage regulators, tap-changing transformers, and thermostats. Restored loads increase the stress on the high-voltage network by increasing the reactive power consumption and causing further voltage reduction. A rundown situation causing voltage instability occurs when load dynamics attempt to restore power consumption beyond the capability of the transmission network and the connected generation.

A major factor contributing to voltage instability is the voltage drop that occurs when active and reactive power flow through inductive reactances of the transmission network; this limits the capability of the transmission network for power transfer and voltage support. The power transfer and voltage support are further limited when some of the generators hit their field or armature current time-overload capability limits.

Voltage stability is threatened when a disturbance increases the reactive power demand beyond the sustainable capacity of the available reactive power resources. Although the most common form of voltage instability is the progressive drop of bus voltages, the risk of overvoltage instability also exists and has been experienced at least on one system. It is caused by a capacitive behavior of the network (extra high voltage transmission lines operating below surge impedance loading) as well as by underexcitation limiters preventing generators or synchronous compensators from absorbing the excess reactive power. In this case, the instability is associated with the inability of the combined generation and transmission system to operate below some load level. In their attempt to restore this load power, transformer tap changers cause long-term voltage instability.

Voltage stability problems may also be experienced at the terminals of high-voltage direct current (HVDC) links used for either long distance or back-to-back applications. They are usually associated with HVDC links connected to weak AC systems and may occur at rectifier or inverter stations; they are also associated with the unfavorable reactive power "load" characteristics of the converters. The HVDC link control strategies significantly influence such problems, since the active and reactive power at the AC–DC junction is determined by the controls. If the resulting loading on the AC transmission stresses it beyond its capability, voltage instability occurs. Such a phenomenon is relatively fast with the time frame of interest being on the order of 1 second or less. Voltage instability may also be associated with converter transformer tap-changer controls, which is a considerably slower phenomenon. Recent developments in HVDC technology (voltage source converters and capacitor commutated converters) have significantly increased the limits for stable operation of HVDC links in weak systems compared with the limits for line commutated converters.

One form of voltage stability problem that results in uncontrolled overvoltages is the self-excitation of synchronous machines. This can arise if the capacitive load of a synchronous machine is too large. Examples of excessive capacitive loads that can initiate self-excitation are open-ended high-voltage lines and shunt capacitors and filter banks from HVDC stations. The overvoltages that result when generator load changes to capacitive are characterized by an instantaneous rise at the instant of change followed by a more gradual rise. This latter rise depends on the relation between the capacitive load component and machine reactances together with the excitation system of the synchronous machine. Negative field current capability of the exciter is a feature that has a positive influence on the limits for self-excitation.

As in the case of rotor angle stability, it is useful to classify voltage stability into the following subcategories:

- *Large-disturbance voltage stability* refers to the system's ability to maintain steady voltages following large disturbances such as system faults, loss of generation, or circuit contingencies. This ability is determined by the system and load characteristics and the interactions of both continuous and discrete controls and protections. Determination of large-disturbance voltage stability requires the examination of the nonlinear response of the power system over a period of time sufficient to capture the performance and interactions of such devices as motors, under load transformer tap changers, and generator field-current limiters. The study period of interest may extend from a few seconds to tens of minutes.
- *Small-disturbance voltage stability* refers to the system's ability to maintain steady voltages when subjected to small perturbations such as incremental changes in system load. This form of stability is influenced by the characteristics of loads, continuous controls, and discrete controls at a given instant of time. This concept is useful in determining, at any instant, how the system voltages will respond to small system changes. With appropriate assumptions, system equations can be linearized for analysis thereby allowing computation of valuable sensitivity information useful in identifying factors influencing stability. This linearization, however, cannot account for nonlinear effects such as tap changer controls (deadbands, discrete tap steps, and time delays). Therefore, a combination of linear and nonlinear analyses is used in a complementary manner.

As noted already, the time frame of interest for voltage stability problems may vary from a few seconds to tens of minutes. Therefore, voltage stability may be either a short-term or a long-term phenomenon.

- *Short-term voltage stability* involves dynamics of fast-acting load components such as induction motors, electronically controlled loads, and HVDC converters. The study period of interest is in the order of several seconds, and analysis requires solution of appropriate system differential equations; this is similar to analysis of rotor angle stability. Dynamic modeling of loads is often essential. In contrast to angle stability, short circuits near loads are important. It is recommended that the term *transient voltage stability* not be used.

- *Long-term voltage stability* involves slower-acting equipment such as tap-changing transformers, thermostatically controlled loads, and generator current limiters. The study period of interest may extend to several or many minutes, and long-term simulations are required for analysis of system dynamic performance. Stability is usually determined by the resulting outage of equipment rather than the severity of the initial disturbance. Instability is due to the loss of long-term equilibrium (e.g., when loads try to restore their power beyond the capability of the transmission network and connected generation), postdisturbance steady-state operating point being small-disturbance unstable, or a lack of attraction toward the stable postdisturbance equilibrium (e.g., when a remedial action is applied too late). The disturbance could also be a sustained load buildup (e.g., morning load increase). In many cases, static analysis can be used to estimate stability margins, identify factors influencing stability, and screen a wide range of system conditions and a large number of scenarios. Where timing of control actions is important, this should be complemented by quasi-steady-state time-domain simulations.

3.4.9   Circuit Breakers

A circuit breaker is an automatically operated electrical switch designed to protect an electrical circuit from damage caused by overload or short circuit. Its basic function is to detect a fault condition and, by interrupting continuity, to immediately discontinue electrical flow. Unlike a fuse, which operates once and then has to be replaced, a circuit breaker can be reset (either manually or automatically) to resume normal operation. Circuit breakers are made in varying sizes, from small devices that protect an individual household appliance up to large switchgear designed to protect high-voltage circuits feeding an entire city [2–3].

3.4.9.1   Operation

All circuit breakers have common features in their operation, although details vary substantially depending on the voltage class, current rating, and type of circuit breaker. The circuit breaker must detect a fault condition; in low-voltage circuit breakers this is usually done within the breaker enclosure. Circuit breakers for large currents or high voltages are usually arranged with pilot devices to sense a fault current and to operate the trip-opening mechanism. The trip solenoid that releases the latch is usually energized by a separate battery, although some high-voltage circuit breakers are self-contained with current transformers, protection relays, and an internal control power source.

Once a fault is detected, contacts within the circuit breaker must open to interrupt the circuit; some mechanically stored energy (using something such as springs or compressed air) contained within the breaker is used to separate the contacts, although some of the energy required may be obtained from the fault current itself. Small circuit breakers may be manually operated; larger units have solenoids to trip the mechanism and electric motors to restore energy to the springs.

The circuit breaker contacts must carry the load current without excessive heating and must also withstand the heat of the arc produced when interrupting the circuit. Contacts are made of copper or copper alloys, silver alloys, and other materials. Service life of the contacts is limited by the erosion due to interrupting the arc. Miniature and molded case circuit breakers are usually discarded when the contacts are worn, but power circuit breakers and high-voltage circuit breakers have replaceable contacts.

When a current is interrupted, an arc is generated. This arc must be contained, cooled, and extinguished in a controlled way so that the gap between the contacts can again withstand the voltage in the circuit. Different circuit breakers use vacuum, air, insulating gas, or oil as the medium in which the arc forms. Different techniques are used to extinguish the arc, including:

- Lengthening of the arc
- Intensive cooling (in jet chambers)
- Division into partial arcs
- Zero-point quenching—contacts open at the zero current time crossing of the AC waveform, effectively breaking no load current at the time of opening. The zero crossing occurs at twice the line frequency (i.e., 100 times per second for 50 Hz and 120 times per second for 60 Hz AC)
- Connecting capacitors in parallel with contacts in DC circuits

Finally, once the fault condition has been cleared, the contacts must again be closed to restore power to the interrupted circuit.

3.4.9.2   Arc Interruption

Miniature low-voltage circuit breakers use air alone to extinguish the arc. Larger ratings will have metal plates or nonmetallic arc chutes to divide and cool the arc. Magnetic blowout coils deflect the arc into the arc chute. In larger ratings, oil circuit breakers rely on vaporization of some of the oil to blast a jet of oil through the arc.

Gas (usually sulfur hexafluoride) circuit breakers sometimes stretch the arc using a magnetic field and then rely on the dielectric strength of the sulfur hexafluoride (SF_6) to quench the stretched arc.

Vacuum circuit breakers have minimal arcing (as there is nothing to ionize other than the contact material), so the arc quenches when it is stretched a very small amount (<2–3 mm). Vacuum circuit breakers are frequently used in modern medium-voltage (MV) switchgear to 35,000 volts.

Air circuit breakers may use compressed air to blow out the arc, or, alternatively, the contacts are rapidly swung into a small sealed chamber, the escaping of the displaced air thus blowing out the arc.

Circuit breakers are usually able to terminate all current very quickly: Typically the arc is extinguished between 30 and 150 ms after the mechanism has been tripped, depending on age and construction of the device.

3.4.9.3   Short-Circuit Current

Circuit breakers are rated both by the normal current that they are expected to carry and the maximum short-circuit current that they can safely interrupt.

Under short-circuit conditions, a current many times greater than normal can exist. When electrical contacts open to interrupt a large current, there is a tendency for an arc to form between the opened contacts, which would allow the current to continue. This condition can create conductive ionized gasses and molten or vaporized metal, which can cause further continuation of the arc, or additional short circuits, which can potentially result in the explosion of the circuit breaker and the equipment that it is installed in. Therefore, circuit breakers must incorporate various features to divide and extinguish the arc.

In air-insulated and miniature breakers an arc-chute structure consisting (often) of metal plates or ceramic ridges cools the arc, and magnetic blowout coils deflect the arc into the arc chute. Larger circuit breakers such as those used in electrical power distribution may use a vacuum or an inert gas such as sulfur hexafluoride or may have contacts immersed in oil to suppress the arc.

The maximum short-circuit current that a breaker can interrupt is determined by testing. Application of a breaker in a circuit with a prospective short-circuit current higher than the breaker's interrupting capacity rating may result in failure of the breaker to safely interrupt a fault. In a worst-case scenario the breaker may successfully interrupt the fault, only to explode when reset.

Miniature circuit breakers used to protect control circuits or small appliances may not have sufficient interrupting capacity to use at a panel board; these circuit breakers are called *supplemental circuit protectors* to distinguish them from distribution-type circuit breakers.

3.4.9.4 Types of Circuit Breakers

Many different classifications of circuit breakers can be made, based on their features such as voltage class, construction type, interrupting type, and structural features.

3.4.9.4.1 Low-Voltage Circuit Breakers Low-voltage (less than 1,000 V_{AC}) types are common in domestic, commercial, and industrial application and include the following:

- Miniature circuit breaker (MCB)—rated current not more than 100 A. Trip characteristics normally not adjustable. Thermal or thermal-magnetic operation.
- Molded case circuit breaker (MCCB)—rated current up to 2,500 A. Thermal or thermal-magnetic operation. Trip current may be adjustable in larger ratings.

The characteristics of low-voltage circuit breakers are given by international standards such as International Electro-technical Commission (IEC) 947. These circuit breakers are often installed in draw-out enclosures that allow removal and interchange without dismantling the switchgear.

Large low-voltage molded case and power circuit breakers may have electrical motor operators, allowing them to be tripped (opened) and closed under remote control. These may form part of an automatic transfer switch system for standby power.

Low-voltage circuit breakers are also made for DC applications, such as DC supply for subway lines. Special breakers are required for direct current because the arc does not have a natural tendency to go out on each half-cycle as for alternating current. A direct current circuit breaker will have blow-out coils that generate a magnetic field that rapidly stretches the arc when interrupting direct current. Small circuit breakers are either installed directly in equipment or are arranged in a breaker panel.

3.4.9.4.2 Magnetic Circuit Breaker Magnetic circuit breakers use a solenoid (electromagnet) whose pulling force increases with the current. Certain designs use electromagnetic forces in addition to those of the solenoid. The circuit breaker contacts are held closed by a latch. As the current in the solenoid increases beyond the rating of the circuit breaker, the solenoid's pull releases the latch, which then allows the contacts to open by spring action. Some types of magnetic breakers incorporate a hydraulic time-delay feature using a viscous fluid. The core is restrained by a spring until the current exceeds the breaker rating. During an overload, the speed of the solenoid motion is restricted by the fluid. The delay permits brief current surges beyond normal running current for motor starting and energizing equipment, for example. Short-circuit currents provide sufficient solenoid force to release the latch regardless of core position, thus bypassing the delay feature. Ambient temperature affects the time delay but does not affect the current rating of a magnetic breaker.

3.4.9.4.3 Thermal Magnetic Circuit Breaker Thermal magnetic circuit breakers, which are the type found in most distribution boards, incorporate both techniques with the electromagnet responding instantaneously to large surges in current (short circuits) and the bimetallic strip responding to less extreme but longer-term overcurrent conditions. The thermal portion of the circuit breaker provides an *inverse time* response feature that provides faster or slower response for larger or smaller overcurrents, respectively.

3.4.9.4.4 Common Trip Breakers Three-pole common trip breakers for supplying a three-phase device have a 2 A rating. When supplying a branch circuit with more than one live conductor, each live conductor must be protected by a breaker pole. To ensure that all live conductors are interrupted when any pole trips, a common trip breaker must be used. These either may contain two or three tripping mechanisms within one case or, for small breakers, may externally tie the poles together via their operating handles. Two-pole common trip breakers are common on 120/240 volt systems where 240 volt loads (including major appliances or further distribution boards) span the two live wires. Three-pole common trip breakers are typically used to supply three-phase electric power to large motors or further distribution boards.

Two- and four-pole breakers are used when there is a need to disconnect the neutral wire to be sure that no current can flow back through the neutral wire from other loads connected to the same network when people need to touch the wires for maintenance. Separate circuit breakers must never be used for disconnecting live and neutral, because if the neutral gets disconnected while the live conductor stays connected, a dangerous condition arises: The circuit will appear deenergized (appliances will not work), but wires will stay live and residual-current devices (RCDs) will not trip if someone touches the live wire (because RCDs need power to trip). This is why only common trip breakers must be used when the neutral wire needs to be switched.

3.4.9.4.5 Medium-Voltage Circuit Breakers Medium-voltage circuit breakers rated between 1 and 72 kV may be assembled into metal-enclosed switchgear line ups for indoor use or may be individual components installed outdoors in a substation. Air-break circuit breakers replaced oil-filled units for indoor applications but are now themselves being replaced by vacuum circuit breakers (up to about 35 kV). Like the high-voltage circuit breakers described below, these are also operated by current sensing protective relays operated through current transformers. The characteristics of MV breakers are given by international standards such as IEC 62271. Medium-voltage circuit breakers nearly always use separate current sensors and protective relays, instead of relying on built-in thermal or magnetic overcurrent sensors.

Medium-voltage circuit breakers can be classified by the medium used to extinguish the arc:

- Vacuum circuit breaker—With rated current up to 3,000 A, these breakers interrupt the current by creating and extinguishing the arc in a vacuum container. These are generally applied for voltages up to about 35,000 V, which corresponds roughly to the medium-voltage range of power systems. Vacuum circuit breakers tend to have longer life expectancies between overhaul than do air circuit breakers.
- Air circuit breaker—Rated current up to 10,000 A. Trip characteristics are often fully adjustable including configurable trip thresholds and delays. Usually electronically controlled, though some models are microprocessor controlled via an integral electronic trip unit. Often used for main power distribution in large industrial plant, where the breakers are arranged in draw-out enclosures for ease of maintenance.
- SF$_6$ circuit breakers extinguish the arc in a chamber filled with sulfur hexafluoride gas.

Medium-voltage circuit breakers may be connected into the circuit by bolted connections to bus bars or wires, especially in outdoor switchyards. Medium-voltage circuit breakers in switchgear line-ups are often built with draw-out construction, allowing the breaker to be removed without disturbing the power circuit connections, using a motor-operated or hand-cranked mechanism to separate the breaker from its enclosure.

3.4.9.4.6 High-Voltage Circuit Breakers Electrical power transmission networks are protected and controlled by high-voltage breakers. The definition of high voltage varies but in power transmission work is usually thought to be 72.5 kV or higher, according to a recent definition by the International Electro-technical Commission (IEC). High-voltage breakers are nearly always solenoid operated, with current sensing protective relays operated through current transformers. In substations the protective relay scheme can be complex, protecting equipment and buses from various types of overload or ground–earth fault.

High-voltage breakers are broadly classified by the medium used to extinguish the arc:

- Bulk oil
- Minimum oil
- Air blast
- Vacuum
- SF_6

Due to environmental and cost concerns over insulating oil spills, most new breakers use SF_6 gas to quench the arc. Circuit breakers can be classified as *live tank*, where the enclosure that contains the breaking mechanism is at line potential, or *dead tank*, with the enclosure at earth potential. High-voltage AC circuit breakers are routinely available with ratings up to 765 kV, and 1,200 kV breakers are likely to come into market very soon. High-voltage circuit breakers used on transmission systems may be arranged to allow a single pole of a three-phase line to trip instead of tripping all three poles; for some classes of faults this improves the system stability and availability.

3.4.9.4.7 Sulfur Hexafluoride (SF6) High-Voltage Circuit Breakers A sulfur hexafluoride circuit breaker uses contacts surrounded by sulfur hexafluoride gas to quench the arc. They are most often used for transmission-level voltages and may be incorporated into compact gas-insulated switchgear. In cold climates, supplemental heating or derating of the circuit breakers may be required due to liquefaction of the SF_6 gas.

3.4.9.4.8 Other Breakers There are some other types of breakers as the following:

1. Breakers for protections against earth faults too small to trip an overcurrent device:
    a. Residual-current device (formerly known as a residual-current circuit breaker): Detects current imbalance but does not provide over-current protection.
    b. Residual current breaker with overcurrent protection (RCBO): Combines the functions of an RCD and an MCB in one package. In the United States and Canada, panel-mounted devices that combine ground (earth) fault detection and overcurrent protection are called ground fault circuit interrupter (GFCI) breakers;

a wall-mounted outlet device providing ground fault detection only is called a Ground Fault Interrupter (GFI).
   c.   Earth leakage circuit breaker (ELCB): This detects earth current directly rather than detecting imbalance. It is no longer seen in new installations for various reasons.
2. Autorecloser: A type of circuit breaker that closes again after a delay. These are used on overhead power distribution systems to prevent short duration faults from causing sustained outages.
3. Polyswitch (polyfuse): A small device commonly described as an automatically resetting fuse rather than a circuit breaker.

3.4.10   Power System Control

The flows of active power and reactive power in a transmission network are fairly independent of each other and are influenced by different control actions. Hence, they may be studied separately for a large class of problems. Active power control is closely related to frequency control, and reactive power control is closely related to voltage control. Since constancy of frequency and voltage are important factors in determining the quality of power supply, it is vital to the satisfactory performance of power systems to be able to control active and reactive power [10].

3.4.10.1   Active Power-Frequency Control

For satisfactory operation of a power system, the frequency should remain nearly constant. Relatively close control of frequency ensures constancy of speed of induction and synchronous motors. Constancy of speed of motor drives is particularly important for satisfactory performance of generating units as they are highly dependent on the performance of all the auxiliary drives associated with the fuel, the feed-water, and the combustion air supply systems. In a network, considerable drop in frequency could result in high magnetizing currents in induction motors and transformers. The extensive use of electric clocks and the use of frequency for other timing purposes require accurate maintenance of synchronous time, which is proportional to integral of frequency. As a consequence, it is necessary to regulate not only the frequency but also its integral.

The frequency of a system is dependent on active power balance. Since frequency is a common factor throughout the system, a change in active power demand at one point is reflected throughout the system by a change in frequency. Because there are many generators supplying power into the system, some means must be provided to allocate change in demand to the generators. A speed governor on each generating unit provides the primary speed control function, whereas supplementary control originating at a central control center allocates generation.

In an interconnected system with two or more independent control areas in addition to control of frequency, the generation within each area has to be controlled to maintain scheduled power interchange. The control of generation and frequency is commonly referred to as load-frequency control (LFC).

3.4.10.2 Reactive Power-Voltage Control

For efficient and reliable operation of power systems, the control of voltage and reactive power should satisfy the following objectives:

1. Voltages at the terminals of all equipment in the system are within acceptable limits. Both utility equipment and customer equipment are designed to operate at a certain voltage rating. Prolonged operation of the equipment at voltages outside the allowable range could adversely affect their performance and possibly cause them damage.
2. System stability is enhanced to maximize use of the transmission system. Voltage and reactive power control have a significant impact on system stability.
3. The reactive power flow is minimized to reduce RI^2 and XI^2 losses to a practical minimum. This ensures that the transmission system operates efficiently (i.e., mainly for active power transfer).

The problem of maintaining voltages within the required limits is complicated by the fact that the power system supplies power to a vast number of loads and is fed from many generating units. Because loads vary, the reactive power requirements of the transmission system vary. This is abundantly clear from the performance characteristics of transmission lines. Since reactive power cannot be transmitted over long distances, voltage control has to be effected by using special devices dispersed throughout the system. This is in contrast to the control of frequency, which depends on the overall system active power balance. The proper selection and coordination of equipment for controlling reactive power and voltage are among the major challenges of power system engineering.

3.4.10.3 Methods of Voltage Control

The control of voltage levels is accomplished by controlling the production, absorption, and flow of reactive power at all levels in the system. The generating units provide the basic means of voltage control; the automatic voltage regulators control field excitation to maintain a scheduled voltage level at the terminals of the generators. Additional means are usually required to control voltage throughout the system. The devices used for this purpose may be classified as follows:

1. Sources or sinks of reactive power, such as shunt capacitors, shunt reactors, synchronous condensers, and static var compensators (SVCs)
2. Line reactance compensators, such as series capacitors
3. Regulating transformers, such as tap-changing transformers and boosters

Shunt capacitors and reactors and series capacitors provide passive compensation. They are either permanently connected to the transmission and distribution system or are switched. They contribute to voltage control by modifying the network characteristics.

Synchronous condensers and SVCs provide active compensation; the reactive power absorbed and supplied by them is automatically adjusted to maintain voltages of the buses to which they are connected. Together with the generating units, they establish voltages at specific points in the system. Voltages at other locations in the system are determined by active and reactive power flows through various circuit elements, including the passive compensation devices.

3.4.10.4   Infinite Bus Concept

The infinite bus, by definition, represents a bus with fixed voltage source. The magnitude, frequency, and phase of the voltage are unaltered by changes in the load (output of the generator).

3.5   Power Quality

Power quality is an important aspect of electrical power systems. Power quality simply means quality of electrical service. Utility, equipment supplier/manufacturer, and consumer are related to power quality issues. A power quality problem may be defined as any power problems manifested in voltage, current, or frequency deviations which results in failure or misoperation of customer equipment. It mostly covers the studies of voltage sags and interruptions, transient overvoltages, fundamentals of harmonics, applied harmonics, long-duration voltage variations, interconnectivity of distributed generation into existing grids, and wiring and grounding.

Power quality can generally be divided into several subcategories, although one category does not necessarily exclude the other. One type of power quality issue is transients. Impulsive transients are sudden undesirable unidirectional increase or decrease in steady-state value of voltage or current usually characterized by rise and decay times. An oscillatory transient, however, is bidirectional and oscillates between positive and negative value of the transient. Another kind of power quality concern is long-duration voltage variations. These can encompass overvoltage and undervoltage in addition to sustained interruptions. Short-duration voltage variations are also of concern. Short-duration interruptions are often caused by faults or equipment failures. Short-duration sags or dips can also be the result of faults on the system or large load increases. A swell is similar to the sag or dip but is a short-duration increase in voltage or current. Phase unbalance is also a concern for power quality. Unbalance can be defined by use of symmetrical components in which the ratio of zero-sequence components to positive-sequence components defines the amount of voltage unbalance. Voltage fluctuations occur when there is a systematic variation of voltage or a series of random changes that do not exceed normal operating ranges. The term *voltage flicker* is often used to describe the variations. Another aspect of the power system subject to power quality problems is system frequency. Frequency variations are any deviation from the nominal frequency of the power system.

Waveform distortion is also of great importance to the power quality of a system. There are many causes of waveform distortion, but they can be classified into five main types: DC offset, harmonics, interharmonics, notching, and noise. DC offset can occur when power converters or other power electronics are subject to a disturbance or some other malfunction that causes asymmetry in the switching. One of the main concerns with DC offset is the saturation of magnetic transformer cores. Harmonics are sinusoidal components of the waveform that are integer multiples of the fundamental frequency of the waveform. Harmonic content is typically described by the total harmonic distortion. Interharmonics, however, are similar to harmonics but occur at noninteger multiples of the fundamental frequency. Principal causes of interharmonics include, but are not limited to, converters, induction furnaces, and arcing devices that can excite resonances in the system. Notching is also a concern. Notching is a periodic distortion in the waveform that is caused generally by switching power converters. Finally, noise can be defined as broadband distortion of the waveform. It is any undesirable distortion that cannot be classified as harmonics or transients.

Current state of power quality research is focused on improving the power quality in both large and small electric grid systems through the use of next-generation power electronics. The two major problems being considered are, first, how to improve power quality during and after fault conditions, and, second, how to improve system robustness and power quality issues when distributed generators go online or offline. With the rise of noncentralized power generation as well as microgrids in ships and other vehicles, maintaining power quality when the number and location of the grid generators is changing has become increasingly important. Also, since more and more power electronics are being used in systems to maintain power quality and perform other tasks, the amount of unwanted harmonics has become a major issue in recent years, especially in the case of triplen harmonics, which can cause major problems in a power system by overloading the neutral line and, if they are higher frequency, injecting noise into other systems (e.g., telephone lines) that may be nearby. Another issue that is being researched heavily is the problem of optimally distributing power quality monitors in a system and how to use those monitors to locate the source of a power quality disruption event.

3.6 Chapter Summary

This chapter deals with the basics of DC machines, synchronous machines, induction machines, transformer, circuit breaker, power system transmission and distribution, power system control, and electrical power systems quality. Induction generators, wound–field synchronous generators, permanent magnet synchronous generators, and synchronous reluctance generators are mostly used as wind generators. Their principles of operation, fundamental characteristics, and parameters are discussed in this chapter.

References

1. A. E. Fitzgerald, C. Kingsley Jr., and S. D. Umans, *Electric machinery*, McGraw-Hill, 1985.
2. P. Kundur, *Power system stability and control*, McGraw-Hill, Inc., 1994.
3. K. R. Padiyar, *Power system dynamics: Stability and control*, John Wiley & Sons (Asia) Pte Ltd, 1996.
4. J. J. Grainger and W. D. Stevenson Jr., *Power system analysis*, McGraw-Hill, Inc., 1994.
5. D. E. Cameron and J. H. Lang, "The control of high-speed variable reluctance generators in electric power systems," *IEEE Trans. Ind. Appl.*, vol. 29, no. 6, pp. 1106–1109, November–December 1993.
6. C. Ferreira, S. R. Jones, W. Heglund, and W. D. Jones, "Detailed design of a 30-kw switched reluctance starter/generator system for a gas turbine engine applications," *IEEE Trans. Ind. Appl.*, vol. 31, no. 3, pp. 553–561, May–June 1995.
7. Y. Sozer and D. Torrey, "Closed loop control of excitation parameters for high speed switched-reluctance generators," *IEEE Trans. Power Electron.*, vol. 19, no. 2, pp. 355–362, March 2004.
8. R. Inerka and R. W. A. A. De Doncker, "High-dynamic direct average torque control for switched reluctance machines," *IEEE Trans. Ind. Appl.*, vol. 39, no. 4, pp. 1040–1045, July–August 2003.
9. P. Kundur, J. Paserba, V. Ajjarapu, G. Anderson, A. Bose, C. Canizares, et al., "Definition and classification of power system stability," *IEEE Trans. Power Systems*, vol. 19, no. 2, pp. 1387–1401, May 2004.
10. P. M. Anderson and A. A. Fouad, *Power system control and stability*, IEEE Press, 1994.
11. IEEE TF Report, "Proposed terms and definitions for power system stability," *IEEE Trans. Power Apparatus and Systems*, vol. PAS-101, pp. 1894–1897, July 1982.

12. T. Van Cutsem, "Voltage instability: Phenomenon, countermeasures and analysis methods," *Proc. IEEE*, vol. 88, pp. 208–227, 2000.
13. D. J. Hill, "Nonlinear dynamic load models with recovery for voltage stability studies," *IEEE Trans. Power Systems*, vol. 8, pp. 166–176, February 1993.
14. T. Van Cutsem and R. Mailhot, "Validation of a fast voltage stability analysis method on the Hydro-Quebec System," *IEEE Trans. Power Systems*, vol. 12, pp. 282–292, February 1997.
15. G. K. Morison, B. Gao, and P. Kundur, "Voltage stability analysis using static and dynamic approaches," *IEEE Trans. Power Systems*, vol. 8, pp. 1159–1171, August 1993.
16. B. Gao, G. K. Morison, and P. Kundur, "Toward the development of a systematic approach for voltage stability assessment of large-scale power systems," *IEEE Trans. Power Systems*, vol. 11, pp. 1314–1324, August 1996.
17. P. A. Lof, T. Smed, G. Andersson, and D. J. Hill, "Fast calculation of a voltage stability index," *IEEE Trans. Power Systems*, vol. 7, pp. 54–64, February 1992.
18. P. Kundur, D. C. Lee, J. P. Bayne, and P. L. Dandeno, "Impact of turbine generator controls on unit performance under system disturbance conditions," *IEEE Trans. Power Apparatus and Systems*, vol. PAS-104, pp. 1262–1267, June 1985.
19. Q. B. Chow, P. Kundur, P. N. Acchione, and B. Lautsch, "Improving nuclear generating station response for electrical grid islanding," *IEEE Trans. Energy Conversion*, vol. EC-4, pp. 406–413, September 1989.
20. P. Kundur, "A survey of utility experiences with power plant response during partial load rejections and system disturbances," *IEEE Trans. Power Apparatus and Systems*, vol. PAS-100, pp. 2471–2475, May 1981.
21. IEEE Committee Report, "Guidelines for enhancing power plant response to partial load rejections," *IEEE Trans. Power Apparatus and Systems*, vol. PAS-102, pp. 1501–1504, June 1983.
22. IEEE Working Group Report, "Reliability indices for use in bulk power system supply adequacy evaluation," *IEEE Trans. Power Apparatus and Systems*, vol. PAS-97, pp. 1097–1103, July–August 1978.
23. D. J. Hill and I. M.Y. Mareels, "Stability theory for differential/algebraic systems with applications to power systems," *IEEE Trans. Circuits and Systems*, vol. 37, pp. 1416–1423, November 1990.
24. W. R. Vitacco and A. N. Michel, "Qualitative analysis of interconnected dynamical systems with algebraic loops: Well-posedness and stability," *IEEE Trans. Circuits and Systems*, vol. CAS-24, pp. 625–637, November 1977.
25. K. L. Praprost and K. A. Loparo, "A stability theory for constrained dynamical systems with applications to electric power systems," *IEEE Trans. Automatic Control*, vol. 41, pp. 1605–1617, November 1996.
26. S. R. Sanders, J. M. Noworolski, X. Z. Liu, and G. C. Verghese, "Generalized averaging method for power conversion circuits," *IEEE Trans. Power Electronics*, vol. 6, pp. 251–259, April 1991.
27. A. M. Stankovic and T. Aydin, "Analysis of unbalanced power system faults using dynamic phasors," *IEEE Trans. Power Systems*, vol. 15, pp. 1062–1068, July 2000.
28. J. S. Thorp, C. E. Seyler, and A. G. Phadke, "Electromechanical wave propagation in large electric power systems," *IEEE Trans. Circuits and Systems—I: Fundamental Theory and Applications*, vol. 45, pp. 614–622, June 1998.

29. R. A. DeCarlo, M. S. Branicky, S. Pettersson, and B. Lennartson, "Perspectives and results on the stability and stabilizability of hybrid systems," *Proc. IEEE*, vol. 88, no. 7, pp. 1069–1082, July 2000.

30. T. S. Lee and S. Ghosh, "The concept of stability in asynchronous distributed decision-making systems," *IEEE Trans. Systems, Man, and Cybernetics—B: Cybernetics*, vol. 30, pp. 549–561, August 2000.

31. D. J. Hill and P. J. Moylan, "Connections between finite gain and asymptotic stability," *IEEE Trans. Automatic Control*, vol. AC-25, pp. 931–936, October 1980.

32. F. Paganini, "A set-based approach for white noise modeling," *IEEE Trans. on Automatic Control*, vol. 41, pp. 1453–1465, October 1996.

33. L. van der Sluis, W. R. Rutgers, and C. G. A. Koreman, "A physical arc model for the simulation of current zero behavior of high-voltage circuit breakers," *IEEE Trans. Power Del.*, vol. 7, no. 2, pp. 1016–1022, April 1992.

34. J. Kosmac and P. Zunko, "A statistical vacuum circuit breaker model for simulation of transients overvoltages," *IEEE Trans. Power Del.*, vol. 10, no. 1, pp. 294–300, January 1995.

35. M. T. Glinkowski, M. R. Gutierrez, and D. Braun, "Voltage escalation and reignition behavior of vacuum circuit breakers during load shedding," *IEEE Trans. Power Del.*, vol. 12, no. 1, pp. 219–226, January 1997.

36. Q. Bui-Van, B. Khodabakhchian, M. Landry, J. Mahseredjian, and J. Mainville, "Performance of series-compensated line circuit breakers under delayed current-zero," *IEEE Trans. Power Del.*, vol. 12, no. 1, pp. 227–233, January 1997.

37. G. St-Jean, M. Landry, M. Leclerc, and A. Chenier, "A new concept in post-arc analysis to power circuit breakers," *IEEE Trans. Power Del.*, vol. 3, no. 3, pp. 1036–1044, July 1988.

38. G. St-Jean and R. F. Wang, "Equivalence between direct and synthetic short-circuit interruption tests on HV circuit breakers," *IEEE Trans. Power App. Syst.*, vol. PAS-102, no. 7, pp. 2216–2223, July 1983.

39. L. Blahous, "Derivation of circuit breaker parameters by means of Gaussian approximation," *IEEE Trans. Power App. Syst.*, vol. PAS-101, no. 12, pp. 4611–4616, December 1982.

40. E. Thuries, P. Van Doan, J. Dayet, and B. Joyeux-Bouillon, "Synthetic testing method for generator circuit breakers," *IEEE Trans. Power Del.*, vol. PWRD-1, no. 1, pp. 179–184, January 1986.

41. B. Kulicke and H. H. Schramm, "Clearance of short-circuits with delayed current zeros in the Itaipu 500 kV substation," *IEEE Trans. Power App. Syst.*, vol. 99, no. 4, pp. 1406–1414, July–August 1980.

42. J. C. Gómez and M. M. Morcos, "Coordination of voltage sag and overcurrent protection in DG systems," *IEEE Trans. Power Delivery*, vol. 20, no. 4, pp. 214–218, January 2005.

43. D. V. Hertem, M. Didden, J. Driesen, and R. Belmans, "Choosing the correct mitigation method against voltage dips and interruptions: A customer-based approach," *IEEE Trans. Power Delivery*, vol. 22, no. 1, pp. 331–339, January 2007.

44. J. V. Milanovic and Y. Zhang, "Global minimization of financial losses due to voltage sags with FACTS based devices," *IEEE Trans. Power Delivery*, vol. 25, no. 1, pp. 298–306, January 2010.

45. C. Fitzer, M. Barnes, and P. Green, "Voltage sag detection technique for a dynamic voltage restorer," *IEEE Trans. Industry Applications*, vol. 40, no. 1, pp. 203–212, January–February 2004.

46. P. Koner and G. Ledwich, "SRAT-distribution voltage sags and reliability assessment tool," *IEEE Trans. Power Delivery*, vol. 19, no. 2, pp. 738–744, April 2004.

47. M. Popov, L. van der Sluis, G. C. Paap, and H. De Herdt, "Computation of very fast transient overvoltages in transformer windings," *IEEE Trans. Power Delivery*, vol. 18, no. 4, pp. 1268–1274, October 2003.

48. J. Takami, S. Okabe, and E. Zaima, "Study of lighting surge overvoltages at substations due to direct lighting strokes to phase conductors," *IEEE Trans. Power Delivery*, vol. 25, no. 1, pp. 425–433, January 2010.

49. B. Gustavsen, "Study of Transformer Resonant Overvoltages Caused by Cable-Transformer High Frequency Interaction," *IEEE Trans. Power Delivery*, vol. 25, no. 2, pp. 770–779, April 2010.

50. P. Ying and R. Jiangjun, "Investigation of very fast transient overvoltage distribution in taper winding of tesla transformer," *IEEE Trans. Magnetics*, vol. 42, no. 3, pp. 434–441, March 2006.

51. D. Paul and V. Haddadian, "Transient overvoltage protection of shore-to-ship power supply system," *IEEE Trans. Industry Applications*, vol. 47, no. 3, pp. 1193–1200, May–June 2011.

52. P. Giridhar Kini and R. C. Bansal, "Effect of voltage and load variations on efficiencies of a motor-pump system," *IEEE Trans. Energy Conversion*, vol. 25, no. 2, pp. 287–292, June 2010.

53. C.-S. Lam, M.-C. Wong, and Y.-D. Han, "Voltage swell and overvoltage compensation with unidirectional power flow controlled dynamic voltage restorer," *IEEE Trans. Power Delivery*, vol. 23, no. 4, pp. 2513–2521, October 2008.

54. L. Fabricio Auler and R. d'Amore, "Power quality monitoring controlled through low-cost modules," *IEEE Trans. Instrumentation and Measurement*, vol. 58, no. 3, pp. 557–562, March 2009.

55. B. Byman, T. Yarborough, R. Schnorr Von Carolsfeld, and J. Van Gorp, "Using distributed power quality monitoring for better electrical system management," *IEEE Trans. Industry Applications*, vol. 36, no. 5, pp. 1481–1485, September–October 2000.

56. A. Dell'Aquila, M. Marinelli, V. G. Monopoli, and P. Zanchetta, "New power-quality assessment criteria for supply systems under unbalanced and nonsinusoidal conditions," *IEEE Trans. Power Delivery*, vol. 19, no. 3, pp. 1284–1290, July 2004.

57. M. A. Eldery, E. F. El-Saadany, M. M. A. Salama, and A. Vannelli,, "A novel power quality monitoring allocation algorithm," *IEEE Trans. Power Delivery*, vol. 21, no. 2, pp. 768–777, April 2006.

chapter 4

Power Electronics

4.1 Introduction

Power electronics is the application of solid-state electronics for the control and conversion of electric power. Power electronic converters can be found wherever there is a need to modify a form of electrical energy (i.e., change its voltage, current, or frequency). Variable-speed wind generator systems need a power electronic interface between the generator and the grid. Even for fixed-speed wind generator systems, when the energy storage system is connected to the grid power electronics devices are essential. This chapter provides a brief overview and basic understanding of various power electronics devices: rectifiers, inverters, direct current (DC)-to-DC choppers, cycloconverters, pulse width modulation (PWM)-based voltage source converters (VSCs), and current source inverters (CSIs). For detailed study, readers are referred to books of power electronics [1–5].

4.2 Power Devices

Power semiconductor devices are semiconductor devices used as switches in power electronic circuits (e.g., switch-mode power supplies). They are also called power devices or, when used in integrated circuits (ICs), power ICs. Most power semiconductor devices are used only in commutation mode (i.e., either on or off) and are therefore optimized for this. Most of them should not be used in linear operation [1].

Power semiconductor devices first appeared in 1952 with the introduction of the power diode by R. N. Hall. It was made of germanium and had a voltage capability of 200 volts and a current rating of 35 amperes. The thyristor appeared in 1957. Thyristors are able to withstand very high reverse breakdown voltage and are also capable of carrying high current. One disadvantage of the thyristor for switching circuits is that once it is "latched on" in the conducting state it cannot be turned off by external control. The thyristor turn-off is passive; that is, the power must be disconnected from the device. The first bipolar transistors devices with substantial power handling capabilities were introduced in the 1960s. These components overcame some limitations of the thyristors because they can be turned on or off with an applied signal. With the improvements brought about by metal–oxide–semiconductor technology (initially developed to produce integrated circuits), power metal–oxide–semiconductor field-effect transistors (MOSFETs) became available in the late 1970s. International Rectifier introduced a 25 A, 400 V power MOSFET in 1978. These devices allow operation at higher frequency than bipolar transistors but are limited to low-voltage applications. The insulated gate bipolar transistor (IGBT) developed in the 1980s became widely available in the 1990s. This component has the power handling capability of the bipolar transistor, with the advantages of the isolated gate drive of the power MOSFET.

Some common power devices are the power diode, thyristor, power MOSFET, and IGBT. A power diode or MOSFET operates on similar principles to its low-power counterpart but is able to carry a larger amount of current and typically is able to support a larger reverse-bias voltage in the off state. Structural changes are often made in power devices to accommodate the higher current density, higher power dissipation, or higher reverse breakdown voltage. The vast majority of discrete (i.e., non-integrated) power devices are built using a vertical structure, whereas small-signal devices employ a lateral structure. With the vertical structure, the current rating of the device is proportional to its area, and the voltage-blocking capability is achieved in the height of the die. With this structure, one of the connections of the device is located on the bottom of the semiconductor die.

4.3   Rectifier

A rectifier is an electrical device that converts alternating current (AC), which periodically reverses direction, to direct current, which is in only one direction, a process known as rectification. Rectifiers have many uses including as components of power supplies and as detectors of radio signals. Rectifiers may be made of solid-state diodes, vacuum tube diodes, mercury arc valves, and other components. When only one diode is used to rectify AC (by blocking the negative or positive portion of the waveform), the difference between the term diode and the term rectifier is merely one of usage—that is, the term rectifier describes a diode that is being used to convert AC to DC. Almost all rectifiers comprise a number of diodes in a specific arrangement for more efficiently converting AC to DC than is possible with only one diode. Before the development of silicon semiconductor rectifiers, vacuum tube diodes and copper(I) oxide or selenium rectifier stacks were used. Figures 4.1, 4.2, and 4.3 show ordinary thyristor-based basic half-wave rectifier, full-wave rectifier, and six-pulse rectifier circuits and their associated output voltage waveforms, respectively.

4.4   Inverter

An inverter is an electrical device that converts DC to AC; the converted AC can be at any required voltage and frequency with the use

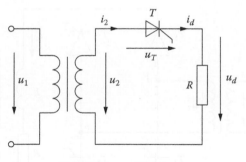

(a) Half-wave rectifier circuit

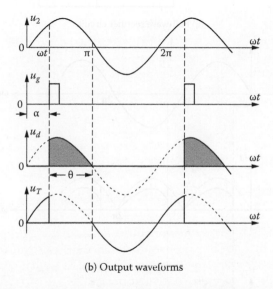

(b) Output waveforms

Figure 4.1 Half-wave rectifier circuit and output waveforms.

of appropriate transformers, switching, and control circuits. Solid-state inverters have no moving parts and are used in a wide range of applications, from small switching power supplies in computers to large electric utility high-voltage direct current applications that transport bulk power. Inverters are commonly used to supply AC power from DC sources such as solar panels or batteries.

There are two main types of inverters. The output of a modified sine wave inverter is similar to a square wave output except that the output goes to zero volts for a time before switching positive or negative. It is simple and low cost and is compatible with most electronic devices, except for sensitive or specialized equipment, for example, certain laser printers. A pure sine wave inverter produces a nearly perfect

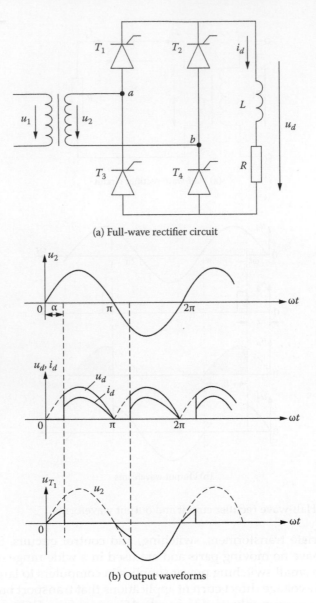

(a) Full-wave rectifier circuit

(b) Output waveforms

Figure 4.2 Full-wave rectifier circuit and output waveforms.

sine wave output (<3% total harmonic distortion) that is essentially the same as utility-supplied grid power. Thus, it is compatible with all AC electronic devices. This is the type used in grid-tie inverters. Its design is more complex and costs 5 to 10 times more per unit power. The electrical inverter is a high-power electronic oscillator. It is so named because

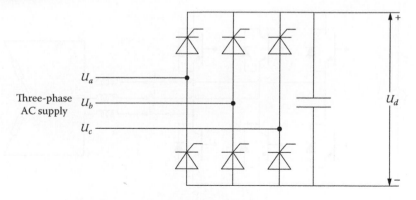

(a) Six-pulse rectifier circuit

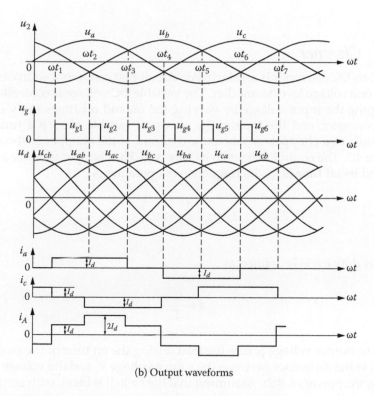

(b) Output waveforms

Figure 4.3 Six-pulse rectifier circuit and output waveforms.

early mechanical AC-to-DC converters were made to work in reverse and thus were "inverted" to convert DC to AC. Figure 4.4 shows a basic inverter circuit.

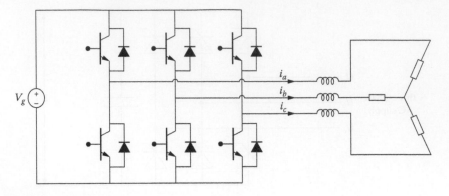

Figure 4.4 Basic inverter circuit.

4.5 Chopper

A DC-to-DC converter is an electronic circuit that converts a source of DC from one voltage level to another. The variable DC voltage is controlled by chopping the input voltage by varying the on and off times (duty cycle) of a converter, and the type of the converter capable of such a function is known as a chopper. A schematic diagram of the chopper is shown in Figure 4.5. The control voltage to its gate is v_c. The chopper is on for a time t_{on}, and its off time is t_{off}. Its frequency of operation is

$$f_c = \frac{1}{(t_{on} + t_{off})} = \frac{1}{T}$$

and its duty cycle is defined as

$$d = \frac{t_{on}}{T}$$

The output voltage across the load during the on time of the switch is equal to the difference between the source voltage V_s and the voltage drop across the power switch. Assuming that the switch is ideal, with zero voltage drop, the average output voltage V_{dc} is given as

$$V_{dc} = \frac{t_{on}}{T} V_s = d V_s$$

where V_s is the source voltage.

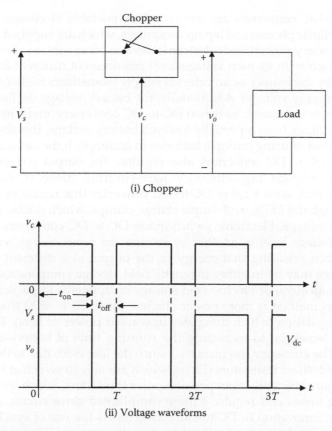

(i) Chopper

(ii) Voltage waveforms

Figure 4.5 Chopper schematic and its waveforms.

Varying the duty cycle changes the output voltage. Note that the output voltage follows the control voltage, as shown in Figure 4.5, signifying that the chopper is the voltage amplifier. The duty cycle *d* can be changed in two ways [2]:

1. By keeping the switching or chopping frequency constant and varying the on time to get a changing duty cycle.
2. Keeping the on time constant and varying the chopping frequency to obtain various values of the duty cycle.

A constant switching frequency has the advantages of predetermined switching losses of the chopper, enabling optimal design of the cooling for the power circuit, and predetermined harmonic contents, leading to an optimal input filter. Both of these advantages are lost by varying the switching frequency of the chopper; hence, this technique for chopper control is not prevalent in practice.

DC-to-DC converters are important in portable electronic devices such as cellular phones and laptop computers, which are supplied primarily with battery power. Such electronic devices often contain several subcircuits, each with its own voltage level requirement different from that supplied by the battery or an external supply (sometimes higher or lower than the supply voltage). Additionally, the battery voltage declines as its stored power is drained. Switched DC-to-DC converters offer a method to increase voltage from a partially lowered battery voltage, thereby saving space instead of using multiple batteries to accomplish the same thing.

Most DC-to-DC converters also regulate the output voltage. Some exceptions include high-efficiency light-emitting diode (LED) power sources, which are a kind of DC-to-DC converter that regulates the current through the LEDs, and simple charge pumps, which double or triple the input voltage. Electronic switch-mode DC-to-DC converters convert one DC voltage level to another by storing the input energy temporarily and then releasing that energy to the output at a different voltage. The storage may be in either magnetic field storage components (inductors, transformers) or electric field storage components (capacitors). This conversion method is more power efficient (often 75 to 98%) than linear voltage regulation (which dissipates unwanted power as heat). This efficiency is beneficial to increasing the running time of battery-operated devices. The efficiency has increased since the late 1980s due to the use of power field-effect transistors (FETs), which are able to switch at high frequency more efficiently than power bipolar transistors, which incur more switching losses and require a more complicated drive circuit. Another important innovation in DC-to-DC converters is the use of synchronous rectification replacing the flywheel diode with a power FET with low "on" resistance, thereby reducing switching losses.

Most DC-to-DC converters are designed to move power in only one direction: from the input to the output. However, all switching regulator topologies can be made bidirectional by replacing all diodes with independently controlled active rectification. A bidirectional converter can move power in either direction, which is useful in applications requiring regenerative braking. Drawbacks of switching converters include complexity, electronic noise (i.e., electromagnetic interference [EMI] or radio frequency interference [RFI]), and to some extent cost, although this has come down with advances in chip design.

DC-to-DC converters are now available as integrated circuits needing minimal additional components and also as a complete hybrid circuit component, ready for use within an electronic assembly. A converter where output voltage is lower than the input voltage is called a buck converter. A converter that outputs a voltage higher than the input voltage is called a boost converter. A buck–boost converter provides an output voltage that may be less than or greater than the input voltage (hence the

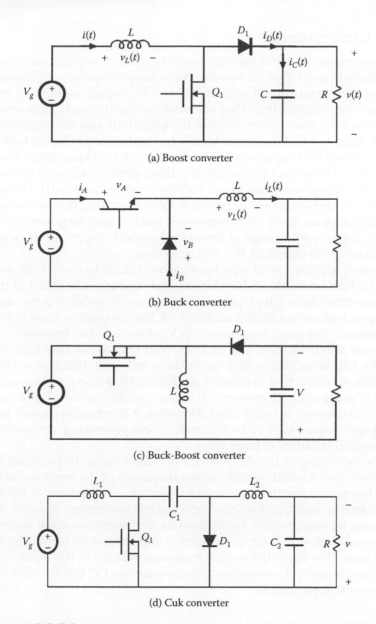

(a) Boost converter

(b) Buck converter

(c) Buck-Boost converter

(d) Cuk converter

Figure 4.6 DC–DC converters.

name buck–boost); the output voltage polarity is opposite to that of the input voltage. Similar to the buck boost converter, the cuk converter provides an output voltage that is less than or greater than the input voltage, but the output voltage polarity is opposite to that of the input voltage. Figure 4.6 shows various chopper circuits.

4.6   Cycloconverter

A cycloconverter or a cycloinverter converts an AC waveform, such as the mains supply, to another AC waveform of a lower frequency, synthesizing the output waveform from segments of the AC supply without an intermediate direct-current link. They are most commonly used in three-phase applications. In most power systems, the amplitude and the frequency of input voltage to a cycloconverter tend to be fixed values, whereas both the amplitude and the frequency of output voltage of a cycloconverter tend to be variable. The output frequency of a three-phase cycloconverter must be less than about one-third to one-half the input frequency [1]. The quality of the output waveform improves if more switching devices are used (a higher pulse number). Cycloverters are used in very large variable frequency drives, with ratings of several megawatts. Figure 4.7 shows the connection of the thyristors in a cycloconverter.

Typical applications of a cycloconverter include to control the speed of an AC traction motor and to start a synchronous motor. Most of these cycloconverters have a high power output—on the order of a few megawatts—and silicon-controlled rectifiers (SCRs) are used in these circuits. By contrast, low-cost, low-power cycloconverters for low-power AC motors are also in use, and many such circuits tend to use TRIACs in place of SCRs. Unlike an SCR, which conducts in only one direction, a TRIAC is capable of conducting in either direction, but it is also a three-terminal device. The use of a cycloconverter is not as common as that of an inverter, and a cycloinverter is rarely used. However, it is common in very high-power applications such as for ball mills in ore processing, cement kilns, and azimuth thrusters in large ships.

The switching of the AC waveform creates noise, or harmonics, in the system that depends mostly on the frequency of the input waveform. These harmonics can damage sensitive electronic equipment. If the relative difference between the input and output waveforms is small, then the converter can produce subharmonics. Subharmonic noise occurs at a frequency below the output frequency and cannot be filtered by load inductance. This limits the output frequency relative to the input. These limitations make cycloconverters often inferior to a DC link converter system for most applications.

4.7   Pulse Width Modulation Scheme

PWM is a commonly used technique for controlling power to inertial electrical devices, made practical by modern electronic power switches. The average value of voltage (and current) fed to the load is controlled by turning the switch between supply and load on and off at a fast pace. The longer the switch is on compared with the off periods, the higher the power

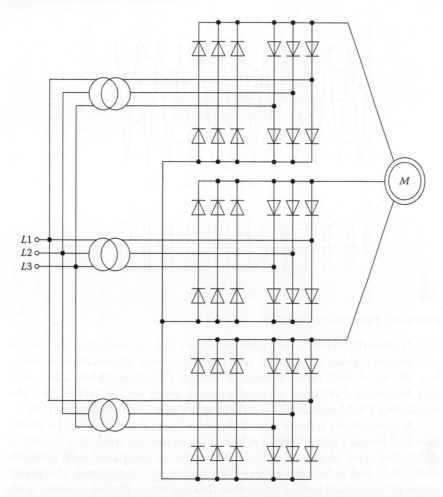

Figure 4.7 Connection of the thyristors in a cycloconverter.

supplied to the load is. The PWM switching frequency has to be much faster than what would affect the load, i.e., the device that uses the power. Typically switchings have to be done several times a minute in an electric stove, 120 Hz in a lamp dimmer, from a few kilohertz (kHz) to tens of kHz for a motor drive, and well into the tens or hundreds of kHz in audio amplifiers and computer power supplies. The term *duty cycle* describes the proportion of on time to the regular interval or period of time; a low duty cycle corresponds to low power, because the power is off for most of the time. Duty cycle is expressed in percent, 100% being fully on. Figure 4.8 shows a figure representing a pulse width modulation scheme.

There are various PWM schemes. Well known among these are sinusoidal PWM, hysteresis PWM, space vector modulation (SVM),

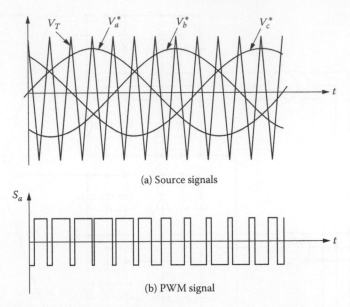

(a) Source signals

(b) PWM signal

Figure 4.8 Pulse-width modulation scheme.

and "optimal" PWM techniques based on the optimization of certain performance criteria, such as selective harmonic elimination, increasing efficiency, and minimization of torque pulsation. While sinusoidal and hysteresis PWM can be implemented using analog techniques, the remaining PWM techniques require the use of a microprocessor [1–5].

A modulation scheme especially developed for drives is the direct flux and torque control (DTC). A two-level hysteresis controller is used to define the error of the stator flux. The torque is compared with its reference value and is fed into a three-level hysteresis comparator. The phase angle of the instantaneous stator flux linkage space phasor together with the torque and flux error state is used in a switching table for the selection of an appropriate voltage state applied to the motor. Usually, there is no fixed pattern modulation in process or fixed voltage-to-frequency relation in the DTC. The DTC approach is similar to the Field Oriented Control (FOC) with hysteresis PWM. However, it takes the interaction among the three phases into account.

The main advantage of PWM is that power loss in the switching devices is very low. When a switch is off there is practically no current, and when it is on there is almost no voltage drop across the switch. Power loss, being the product of voltage and current, is thus close to zero in both cases. PWM also works well with digital controls, which because of their on–off nature can easily set the needed duty cycle. PWM has also been used in certain communication systems where its duty cycle has been used to convey information over a communications channel.

Usually, the on and off states of the power switches in one inverter leg are always opposite. Therefore, the inverter circuit can be simplified into three two-position switches. Either the positive or the negative DC bus voltage is applied to one of the motor phases for a short time. PWM is a method whereby the switched voltage pulses are produced for different output frequencies and voltages. A typical modulator produces an average voltage value equal to the reference voltage within each PWM period. Considering a very short PWM period, the reference voltage is reflected by the fundamental of the switched pulse pattern.

Apart from the fundamental wave, the voltage spectrum at the motor terminals consists of many higher harmonics. The interaction between the fundamental motor flux wave and the fifth and seventh harmonic currents produces a pulsating torque six times that of the fundamental supply frequency. Similarly, the eleventh and thirteenth harmonics produce a pulsating torque twelve times that of the fundamental supply frequency. Furthermore, harmonic currents and skin effect increase copper losses, which leads to motor derating. However, the motor reactance acts as a low-pass filter and substantially reduces high-frequency current harmonics. Therefore, the motor flux (Induction Motor [IM] & Permanent Magnet Synchronous Motor [PMSM]) is in good approximation sinusoidal and the contribution of harmonics to the developed torque is negligible. To minimize the effect of harmonics on the motor performance, the PWM frequency should be as high as possible. However, the PWM frequency is restricted by the control unit (resolution) and the switching device capabilities, for example, due to switching losses and dead time distorting the output voltage.

4.8   PWM VSC

The PWM VSC provides a power electronic interface between the AC power system and superconducting coil. In the PWM generator, the sinusoidal reference signal is phase modulated by means of the phase angle, a, of the VSC output AC voltage. The modulated sinusoidal reference signal is compared with the triangular carrier signal to generate the gate signals for the IGBTs. Figure 4.9 shows a basic PWM-based voltage source converter circuit consisting of a Wye-Delta transformer, a six-pulse PWM rectifier/inverter using an IGBT, and a DC link capacitor [6].

4.9   Current Source Inverter

With the availability of modern gate turn-off switching devices at increasing power levels and the introduction of advanced multilevel power converter topologies, the classical CSI topology has been virtually replaced by the voltage source inverter (VSI), even in applications up to several

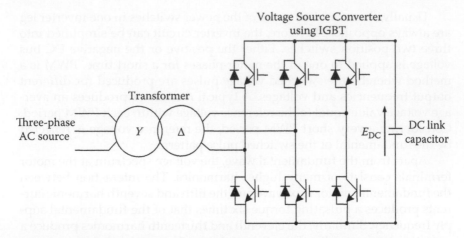

Figure 4.9 Basic configuration of PWM voltage source converter (VSC).

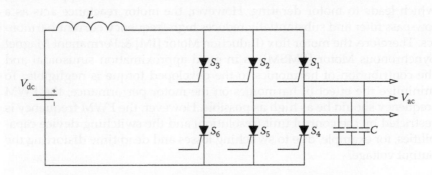

Figure 4.10 Schematic of a classical SCR-based CSI.

MW. However, SCR-based CSI systems are still being used for very high-power synchronous motor drives and utility power systems due to various performance advantages [5].

CSI topologies have certain performance advantages in terms of ruggedness and their ability to feed capacitive and low-impedance loads with ease. As high-power permanent magnet motors with extremely low armature winding inductance are becoming more common and the electrolytic capacitor of a VSI becoming notorious as the largest and least reliable among inverter components, a renewed interest in CSI systems may be expected to follow.

The classical CSI based on SCRs (Figure 4.10) has several disadvantages, stemming simply from the fact that SCRs cannot be turned off from the gate. Hence, their operation has been typically limited to six-step switching and application to active loads capable of operation at leading power factor. Six-step switching leads to a large amount of harmonics in

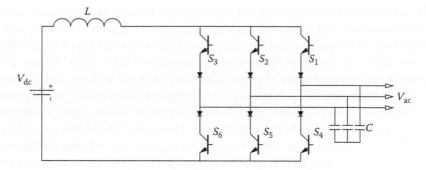

Figure 4.11 Schematic of a CSI using IGBT and series diodes.

the load voltage and current. Thus, they have been naturally bucked by the trend of increasing demands of performance. Furthermore, they are not well suited to drive induction motors, which must operate at a lagging power factor. These reasons have generally impeded their widespread application. Replacement of the SCR with a GTO device would allow turn-off capability and will result in the extension of operation to loads with wider power factor and even PWM capability. However, due the limited switching speeds of GTOs this approach has seen limited application. More commonly, GTO devices have been adapted to operate in multilevel VSI systems. Switching throws in a CSI realized using bidirectional voltage blocking and unidirectional current carrying devices have been well known. Figure 4.11 illustrates such a realization using IGBT devices in series with diodes. This topology is plagued with low efficiency due to current flow through the series connection of two semiconductors per throw.

The three-phase PWM CSI is composed of a bridge with six reverse blocking switches (S1–S6), which each has a transistor and a series diode. The DC link contains an inductor as the main energy storage component, and, at the output, a C (capacitor) filter smoothens the pulsed phase currents from the DC link. The topology is capable of operating at moderate maximum power point (MPP) voltages (ca. 200–400 V) due to its inherent voltage-boost characteristic. However, to keep the volt–second balance across the DC inductor, there is an upper DC voltage limit, which is defined as the absolute minimum of the rectified phase-to-phase voltage.

CSIs, typically supplied from controlled rectifiers with closed-loop current control, can transfer the electric power in both directions and are characterized by a fast response to the phase command for the vector of output current. Operated in the square-wave mode, CSIs are more efficient than the PWM VSIs. Also, the power circuit of the CSI is simpler and more robust than that of the VSI, thanks to the absence of freewheeling diodes and, because of the large DC–link inductance and current control in the rectifier, the inherent protection from overcurrents. On the other hand, the same inductance slows down the response to current magnitude control commands and can cause a dangerous overvoltage if the current path is broken. The square-wave output current waveforms, rich in low-order harmonics, are the most obvious disadvantage of CSIs. In addition, the same waveforms produce voltage spikes in the stator leakage inductance of the supplied motor, potentially dangerous for the winding insulation. PWM CSIs, equipped with output capacitors that shunt ripple currents, offer a partial solution to these problems. However, PWM CSIs have their weaknesses too, such as the increased size and cost, reduced efficiency associated with the PWM operation, greater complexity of the control algorithm, and susceptibility to resonance between the output capacitors and load inductances.

4.10   Chapter Summary

This chapter provides a brief overview of power electronics devices—such as rectifiers, inverters, DC-to-DC choppers, cycloconverters, PWM-based voltage source converters, and current source inverters—which are essential components of grid-connected wind generator systems. In the case of a variable-speed wind generator system, power electronics interface is used between the terminal of the wind generator and the grid point. Again, in the case of the fixed-speed wind generator system, power electronics interface is necessary for the energy storage system. Thus, the basic understanding of the operations and characteristic of the power electronics devices are essential. These concepts will be fully used in Chapters 5, 7, 8, and 9.

References

1. M. H. Rashid, *Power electronics*, 3d ed., Prentice Hall, 2004.
2. R. Krishnan, *Electric motor drives: Modeling, analysis, and control*, Prentice Hall, 2001.
3. N. Mohan, *Electric drives: An integrative approach*, MNPERE, 2000.
4. V. Subrahmanyam, *Electric drives: Concepts and applications*, McGraw-Hill, 1994.
5. B. Wu, *High-power converters and AC drives*, IEEE Press, 2006.
6. M. H. Ali, J. Tamura, and B. Wu, "SMES strategy to minimize frequency fluctuations of wind generator system," *Proceedings of the 34th annual conference of the IEEE Industrial Electronics Society (IECON 2008)*, November 10–13, 2008, Orlando, FL, pp. 3382–3387.
7. V. Blasko and V. Kaura, "A new mathematical model and control of a three-phase AC-DC voltage source converter," *IEEE Trans. Power Electron.*, vol. 12, no. 1, pp. 116–123, January 1997.

8. J. W. Choi and S. K. Sul, "Fast current controller in three-phase AC/DC boost converter using d-q axis cross-coupling," *IEEE Trans. Power Electronics*, vol. 13, no. 1, pp. 179–185, January 1998.

9. R. Wu, S. B. Dewan, and G. R. Slemon, "Analysis of an AC-to-DC voltage source converter using PWM with fixed switching frequency," *IEEE Trans. Ind. Appl.*, vol. 27, no. 2, pp. 355–364, March–April 1991.

10. T. Noguchi, H. Tomiki, S. Kondo, and I. Takahashi, "Direct power control of PWM converter without power-source voltage sensor," *IEEE Trans. Ind. Appl.*, vol. 34, no. 3, pp. 473–479, May–June 1998.

11. M. Malinowski, M. P. Kazmierkowski, S. Hansen, F. Blaabjerg, and G. D. Marques, "Virtual-flux-based direct power control of three-phase PWM rectifier," *IEEE Trans. Ind. Appl.*, vol. 37, no. 4, pp. 1019–1027, July–August 2001.

12. T. Ohnuki, O. Miyashita, P. Lataire, and G. Maggetto, "Control of a three-phase PWM rectifier using estimated AC-side and DC-side voltages," *IEEE Trans. Power Electron.*, vol. 14, no. 2, pp. 222–226, March 1999.

13. J. Rodríguez, S. Bernet, B. Wu, J. Pontt, and S. Kouro, "Multilevel voltage-source-converter topologies for industrial medium-voltage drives," *IEEE Trans. Ind. Electron.*, vol. 54, no. 6, pp. 2930–2945, December 2007.

14. M. Salo and H. Tuusa, "A vector-controlled PWM current-source-inverter fed induction motor drive with a new stator current control method," *IEEE Trans. Ind. Electron.*, vol. 52, no. 2, pp. 523–531, April 2005.

15. P. Cancelliere, V. D. Colli, R. Di Stefano, and F. Marignetti, "Modeling and control of a zero-current-switching DC/AC current-source inverter," *IEEE Trans. Ind. Electron.*, vol. 54, no. 4, pp. 2106–2119, August 2007.

16. M. Hombu, S. Ueda, and A. Ueda, "A current source GTO inverter with sinusoidal inputs and outputs," *IEEE Trans. Ind. Appl.*, vol. IA-23, no. 2, pp. 247–255, March 1987.

17. N. R. Zargari, S. C. Rizzo et al., "A new current-source converter using a symmetric gate-commutated thyristor (SGCT)," *IEEE Trans. Ind. Appl.*, vol. 37, no. 3, pp. 896–903, May–June 2001.

18. S. Rees, "New cascaded control system for current-source rectifiers," *IEEE Trans. Ind. Electron.*, vol. 52, no. 3, pp. 774–784, June 2005.

19. B. M. Han and S. I. Moon, "Static reactive-power compensator using soft switching current-source inverter," *IEEE Trans. Ind. Electron.*, vol. 48, no. 6, pp. 1158–1165, December 2001.

20. M. Salo and H. Tuusa, "A new control system with a control delay compensation for a current-source active power filter," *IEEE Trans. Ind. Electron.*, vol. 52, no. 6, pp. 1616–1624, December 2005.

21. J. R. Espinoza, G. Joos, J. I. Guzman, L. A. Moran, and R. P. Burgos, "Selective harmonic elimination and current/voltage control in current/voltage-source topologies: A unified approach," *IEEE Trans. Ind. Electron.*, vol. 48, no. 1, pp. 71–81, February 2001.

22. J. Ma, B. Wu, and S. Rizzo, "A space vector modulated CSI-based AC drive for multimotor applications," *IEEE Trans. Power Electron.*, vol. 16, no. 4, pp. 535–544, July 2001.

23. Y. Suh, J. K. Steinke, and P. K. Steimer, "Efficiency comparison of voltage-source and current-source drive systems for medium-voltage applications," *IEEE Trans. Ind. Electron.*, vol. 54, no. 5, pp. 2521–2531, October 2007.

24. J. Rodríguez, L. Moran, J. Pontt, R. Osorio, and S. Kouro, "Modeling and analysis of common-mode voltages generated in medium voltage PWMCSI drives," *IEEE Trans. Power Electron.*, vol. 18, no. 3, pp. 873–879, May 2003.

25. R. Emery and J. Eugene, "Harmonic losses in LCI-fed synchronous motors," *IEEE Trans. Ind. Appl.*, vol. 38, no. 4, pp. 948–954, July–August 2002.
26. R. Bhatia, H. Krattiger, A. Bonanini, D. Schafer, J. T. Inge, and G. H. Sydnor, "Adjustable speed drive using a single 135,000 HP synchronous motor," *IEEE Trans. Energy Conversion*, vol. 14, no. 3, pp. 571–576, September 1999.
27. S. Kwak and H. A. Toliyat, "Current-source-rectifier topologies for sinusoidal supply current: Theoretical studies and analyses," *IEEE Trans. Ind. Electron.*, vol. 53, no. 3, pp. 984–987, June 2006.
28. P. Syam, G. Bandyopadhyay, P. K. Nandi, and A. K. Chattopadhyay, "Simulation and experimental study of interharmonic performance of a cycloconverter-fed synchronous motor drive," *IEEE Trans. Energy. Conversion*, vol. 19, no. 2, pp. 325–332, June 2004.
29. Y. Liu, G. T. Heydt, and R. F. Chu, "The power quality impact of cycloconverter control strategies," *IEEE Trans. Power Del.*, vol. 20, no. 2, pp. 1711–1718, April 2005.
30. Z. Wang and Y. Liu, "Modeling and simulation of a cycloconverter drive system for harmonic studies," *IEEE Trans. Ind. Electron.*, vol. 47, no. 3, pp. 533–541, June 2000.
31. F. Zhang, L. Du, F. Z. Peng, and Z. Qian, "A new design method for high-power high-efficiency switched-capacitor DCDC converters," *IEEE Trans. Power Electron.*, vol. 23, pp. 832–840, March 2008.
32. J. Pinheiro and I. Barbi, "The three-level ZVS-PWM DC-to-DC converter," *IEEE Trans. Power Electron.*, vol. 8., pp. 486–492, October 1993.
33. H. Wu and X. He, "Single phase three-level power factor correction circuit with passive lossless snubber," *IEEE Trans. Power Electron.*, vol. 17, pp. 946–953, 2002.
34. J. P. Rodrigues, S. A. Mussa, I. Barbi, and A. J. Perin, "Three-level zero-voltage switching pulse-width modulation DCDC boost converter with active clamping," *IET Power Electronics*, vol. 3, pp. 345–354, 2010.
35. J. C. Rosas-Caro, J. M. Ramirez, F. Z. Peng, and A. Valderrabano, "A DC-DC multilevel boost converter," *IET Power Electronics*, vol. 3, pp. 129–137, 2010.
36. M. Shen, F. Z. Peng, and L. M. Tolbert, "Multilevel DC–DC power conversion system with multiple DC sources," *IEEE Trans. Power Electron.*, vol. 23, pp. 420–426, January 2008.
37. F. H. Khan and L.M. Tolbert, "Multiple-load-source integration in a multilevel modular capacitor-clamped DC–DC converter featuring fault tolerant capability," *IEEE Trans. Power Electron.*, vol. 24, pp. 14–24, January 2009.
38. X. Yuan and I. Barbi, "Fundamentals of a new diode clamping multilevel inverter," *IEEE Trans. Power Electron.*, vol. 15, pp. 711–718, July 2000.
39. F. Z. Peng, "A generalized multilevel inverter topology with self voltage balancing," *IEEE Trans. Ind. Application*, vol. 37, pp. 611–618, March–April 2001.
40. G. P. Adam, S. J. Finney, A. M. Massoud, and B. W. Williams, "Capacitor balance issues of the diode-clamped multilevel inverter operated in a quasi two-state mode," *IEEE Trans. Ind. Electron.*, vol. 55, pp. 3088–3099, 2008.
41. S. Busquets-Monge, S. Alepuz, J. Bordonau, and J. Peracaula, "Voltage balancing control of diode-clamped multilevel converters with passive front-ends," *IEEE Trans. Power Electron.*, vol. 23, pp. 1751–1758, 2008.
42. N. Rouger and J.-C. Crebier, "Toward generic fully integrated gate driver power supplies," *IEEE Trans. Power Electron.*, vol. 23, pp. 2106–2114, July 2008.

chapter 5

Wind Generators

5.1 Introduction

Wind energy conversion systems (WECSs) have become a focal point in the research of renewable energy sources. A wind generator (WG) is a device that generates electrical power from wind energy. Induction generators are mostly used as wind generators. However, synchronous generators can also be used as wind generators. According to rotational speed, there are two types of wind generator systems: (1) fixed speed; and (2) variable speed. This chapter describes the wind generator systems, their characteristics, the maximum power point tracking (MPPT) system, and the method of total efficiency calculation of wind generators.

5.2 Fixed-Speed Wind Energy Conversion Systems

In a fixed-speed WECS, the turbine speed is determined by the grid frequency, the generator pole pairs number, the machine slip, and the gearbox ratio. A change in wind speed will not affect the turbine speed to a large extent but affects the electromagnetic torque and, hence, also the electrical output power. With a fixed-speed WECS, it may be necessary to use aerodynamic control of the blades to optimize the whole system performance, thus introducing additional control systems, complexities, and costs. As for the generating system, nearly all wind turbines installed at present use one of the following systems: squirrel-cage induction generators; doubly fed (wound-rotor) induction generators; or direct-drive synchronous generators. The most used wind turbine systems in this case are illustrated in Figure 5.1. Using induction generators will keep an almost fixed speed (variation of 1–2%). The power is limited aerodynamically by stall, active stall, or pitch control. A soft starter is normally used to reduce the inrush current during startup. A reactive power compensator is also needed to reduce (i.e., almost eliminate) the reactive power demand from the turbine generators. It is usually done by activating continuously the capacitor banks following load variation. Those solutions are attractive due to low cost and high reliability. However, a fixed-speed system cannot extract as much energy from the wind as a variable-speed topology. Today, the variable-speed WECSs are continuously increasing their

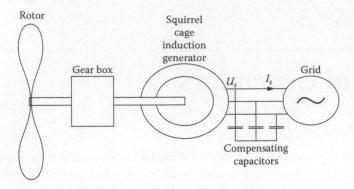

Figure 5.1 Fixed-speed wind energy conversion system.

market share, because it is possible to track the changes in wind speed by adapting shaft speed, thus maintaining optimal energy generation [1].

5.3 Variable-Speed Wind Energy Conversion Systems

The variable-speed generation system is able to store the varying incoming wind power as rotational energy, by changing the speed of the wind turbine. So, the stress on the mechanical structure is reduced and the delivered electrical power becomes smoother. The control system maintains the mechanical power at its rated value by using the MPPT technique. These WECSs are generally divided into two categories: (1) systems with partially rated power electronics; and (2) systems with full-scale power electronic interfacing wind turbines [1].

Figure 5.2 shows two solutions of wind turbines with partially rated power converters. Figure 5.2a shows a WECS with a wound rotor induction generator. Extra resistance controlled by power electronics is added in the rotor, giving a speed range of 2 to 4%. This solution also needs a soft starter and a reactive power compensator. Figure 5.2b shows another solution using a medium-scale power converter with a wound rotor induction generator. In this case, a power converter connected to the rotor through slip rings controls the rotor currents. If the generator is running super synchronously, the electrical power is delivered through both the rotor and the stator. If the generator is running subsynchronously, the electrical power is delivered into the rotor only from the grid. A speed variation of 60% around synchronous speed may be obtained by the use of a power converter of 30% of nominal power. The other WECS category is wind turbines with a full-scale power converter between the generator and grid that gives extra losses in the power conversion, but it will gain the added

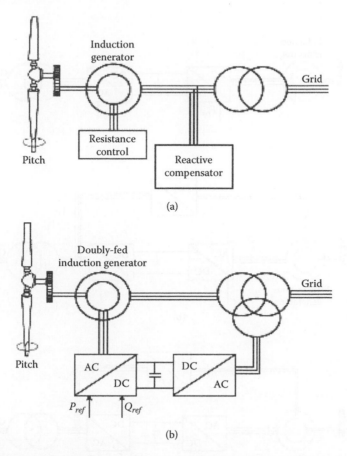

Figure 5.2 Partially rated power electronics WECS: (a) Rotor resistance converter. (b) Doubly fed induction generator.

technical performance. Figure 5.3 shows four possible solutions using an induction generator, a multipole synchronous generator, and a permanent magnet synchronous generator.

5.4 Wind Generators

Most wind turbine manufacturers use six-pole induction (asynchronous) generators, whereas others use directly driven synchronous generators. In the power industry, in general, induction generators are not very common for power production, but induction motors are used worldwide. The power generation industry almost exclusively uses large synchronous generators, as they have the advantage of a variable reactive power production (i.e., voltage control).

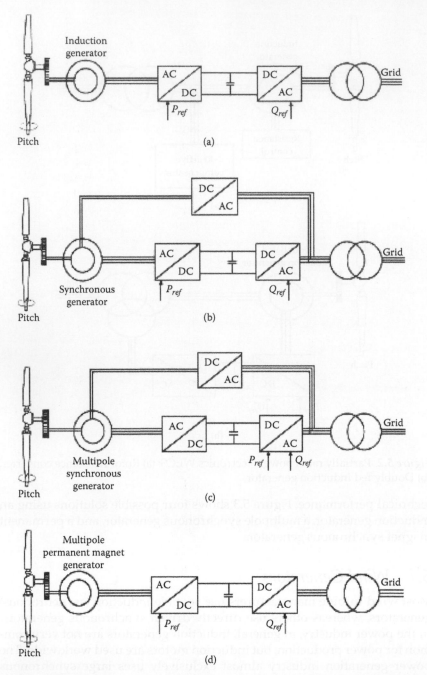

Figure 5.3 Full-scale power electronics WECS: (a) Induction generator. (b) Synchronous generator. (c) Multipole synchronous generator. (d) Multipole permanent magnet generator.

5.4.1 Synchronous Generators

Synchronous generators (SGs) are widely used in stand-alone WECSs where the synchronous generator can be used for reactive power control in the isolated network. To ensure the wind turbine connection to the grid, back-to-back pulse width modulation (PWM) voltage source inverters are interfaced between the synchronous generator and the grid. The grid-side PWM inverter allows for control of real and reactive power transferred to the grid. The generator-side converter is used for electromagnetic torque regulation. Synchronous generators of 500 kW to 2 MW are significantly more expensive than induction generators with a similar size. The use of a multipole synchronous generator (i.e., a large-diameter synchronous ring generator) avoids the installation of a gearbox as an advantage, but there will be a significant increase in weight. Indeed, the industry uses directly driven variable-speed synchronous generators with a large-diameter synchronous ring generator. The variable, directly driven approach avoids the installation of a gearbox, which is essential for medium- and large-scale wind turbines. Employing a permanent magnet synchronous generator provides a solution that is appreciated in small wind turbines, but it cannot be extended to large-scale power because it involves big and heavy permanent magnets.

5.4.2 Induction Generators

Induction generators (IGs) are increasingly used these days because of their relative advantageous features over conventional synchronous generators. These features are brushless and rugged construction, low cost, maintenance and operational simplicity, self-protection against faults, good dynamic response, and capability of generating power at varying speed. The latter feature facilitates the induction generator operation in a stand-alone or isolated mode to supply far-flung and remote areas where grid extension is not economically viable, in conjunction with the synchronous generator to fulfill the increased local power requirement, and in grid-connected mode to supplement the real power demand of the grid by integrating power from resources located at different sites. The reactive power requirements are the disadvantage of induction generators. This reactive power can be supplied by a variety of methods, from simple capacitors to complex power conversion systems.

Induction generators were used for a long time in a constant-speed WECS, where the pitch control or active stall control is dictated for power limitation and protection, a soft starter is also used to limit transients when the generator is connected to the grid. For variable-speed WECSs, back-to-back PWM inverters are used, where the control system of the inverter on the generator side regulates the machine torque and consequently the

Table 5.1 WECS Generator Comparison

Type	Pros	Cons
Induction generator	1. Full-speed range 2. No brushes on the generator 3. Complete control of reactive and active power 4. Proven technology	1. Full-scale power converter 2. Need for gear
Synchronous generator	1. Full-speed range 2. Possibility of avoiding gear 3. Complete control of reactive and active power	1. Small converter for field 2. Full-scale power converter
Permanent magnet synchronous generator	1. Full-speed range 2. Possibility of avoiding gear 3. Complete control of reactive and active power 4. Brushless (low maintenance) 5. No power converter for field	1. Full-scale power converter 2. Multipole generator (big and heavy) 3. Permanent magnets needed
Doubly fed induction generator	1. Limited speed range −30% to 30% around synchronous speed 2. Inexpensive small capacity PWM inverter 3. Complete control of reactive and active power	1. Need slip rings 2. Need for gear

rotor speed, therefore keeping the frequency within defined limits. On the other hand, the inverter on the grid side controls the reactive power at the coupling point. In this case, the doubly fed induction generator is widely used. Indeed, among many variable-speed concepts, WECSs using doubly fed induction generators have many advantages over others. For example, the power converter in such wind turbines deals only with rotor power; therefore, the converter rating can be kept fairly low, approximately 20% of the total machine power. This configuration allows for variable-speed operation while remaining more economical than a series configuration with a fully rated converter. Other features such as the controllability of reactive power help doubly fed induction generators play a similar role to that of synchronous generators.

Table 5.1 briefly gives the pros and cons of the major WECSs detailed in the literature.

5.5   *Wind Generator Characteristics*

In Chapter 2, it is shown that the power captured by the WG blades P_m is a function of the blade shape, the pitch angle, and the radius and the rotor speed of rotation:

$$P_m = \frac{1}{2}\pi\rho C_P(\lambda,\beta)R^2V^3 \qquad (5.1)$$

where ρ is the air density (typically 1.25 kg/m3), β is the pitch angle (in degrees), $C_P(\lambda, \beta)$ is the wind-turbine power coefficient, R is the blade radius (in meters), and V is the wind speed (in m/s). The term λ is the tip–speed ratio, defined as

$$\lambda = \frac{R\Omega}{v} \qquad (5.2)$$

where Ω is the WG rotor speed of rotation (rad/s).

Considering the generator efficiency η_G, the total power produced by WGs, P, is

$$P = \eta_G Pm \qquad (5.3)$$

The WG power coefficient is maximized for a tip–speed ratio value λ_{opt} when the blade's pitch angle is $\beta = 0^0$. The WG power curves for various wind speeds are shown in Figure 5.4. It is observed that, for each wind speed, there exists a specific point in the WG output power versus rotating speed characteristic where the output power is maximized. The control of the WG load results in a variable-speed WG operation, such that maximum power is extracted continuously from the wind (MPPT control). The value of the tip–speed ratio is constant for all maximum power

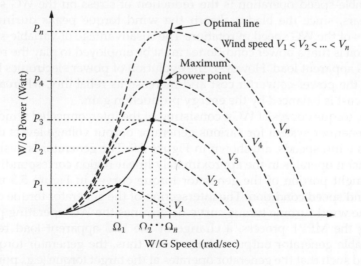

Figure 5.4 WG power curves at various wind speeds.

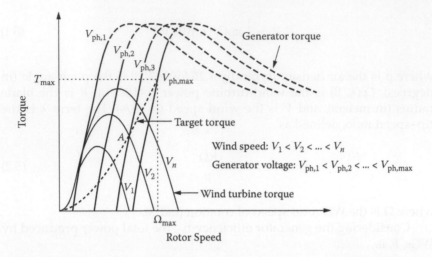

Figure 5.5 Torque-speed characteristics of the wind turbine and the generator.

points (MPPs), whereas the WG speed of rotation is related to the wind speed as follows:

$$\Omega_n = \lambda_{opt} \frac{V_n}{R} \tag{5.4}$$

where Ω_n is the optimal WG speed of rotation at a wind velocity V_n.

Besides the optimal energy production capability, another advantage of variable-speed operation is the reduction of stress on the WG shafts and gears, since the blades absorb the wind torque peaks during the changes of the WG speed of rotation. The disadvantage of variable-speed operation is that a power conditioner must be employed to play the role of the WG apparent load. However, the evolution of power electronics helps reduce the power-converter cost and increase its reliability, whereas the higher cost is balanced by the energy production gain.

The torque curves of WGs, consisting of the interconnected wind-turbine–generator system for various generator output voltage levels under various wind speeds, are shown in Figure 5.5. The generator is designed such that it operates in the approximately linear region corresponding to the straight portion of the generator torque curves in Figure 5.5 under any wind speed condition. The intersection of the generator torque curve with the wind turbine torque curve determines the WG operating point. During the MPPT process, a change of the WG apparent load results in variable generator output voltage level; thus, the generator torque is adjusted such that the generator operates at the target torque (e.g., point A) under any wind speed. The target torque line corresponds to the optimal power production line indicated in Figure 5.4, where the energy extracted from the WG system is maximized.

5.6 Maximum Power Point Tracking System

This section describes the MPPT system in the case of variable-speed wind generator systems. A commonly used WG control system is shown in Figure 5.6a. This topology is based on the WG optimal power versus the rotating speed characteristic, which is usually stored in a microcontroller

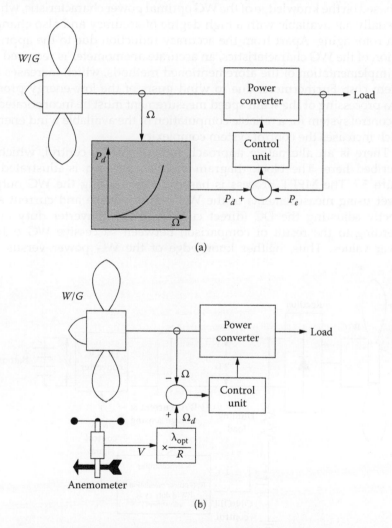

Figure 5.6 WG MPPT methods. (a) Control system based on rotating-speed measurements. (b) Control system based on wind speed measurements.

memory. The WG rotating speed is measured; then, the optimal output power is calculated and compared with the actual WG output power. The resulting error is used to control a power interface [2].

Another control system based on wind speed measurements is shown in Figure 5.6b. The wind speed is measured, and the required rotor speed for maximum power generation is computed. The rotor speed is also measured and compared with the calculated optimal rotor speed, whereas the resulting error is used to control a power interface.

The disadvantage of all the previously outlined methods is that they are based on the knowledge of the WG optimal power characteristic, which is usually not available with a high degree of accuracy and also changes with rotor aging. Apart from the accuracy reduction due to the approximation of the WG characteristics, an accurate anemometer is required for the implementation of the aforementioned methods, which increases the system cost. Furthermore, due to wind gusts of the low-energy profile, extra processing of the wind speed measurement must be incorporated in the control system for a reliable computation of the available wind energy, which increases the control system complexity.

There is an alternative approach for WG MPPT control, which is described here. The block diagram of this approach is illustrated in Figure 5.7. The MPPT process is based on monitoring the WG output power using measurements of the WG output voltage and current and directly adjusting the DC (direct current)-to-DC converter duty cycle according to the result of comparison between successive WG output power values. Thus, neither knowledge of the WG power versus the

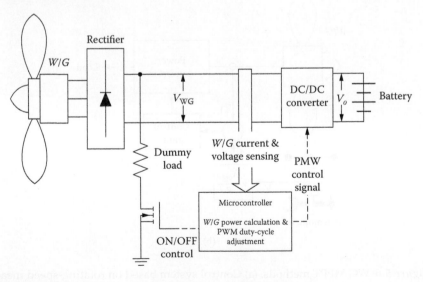

Figure 5.7 Block diagram of the alternative MPPT system.

rotor speed of rotation or wind speed characteristic nor measurements of the wind speed are required. A resistive dummy load is used to protect WGs from overspeeding. This MPPT method does not depend on WG wind and rotor speed ratings or the DC-to-DC converter power rating. Although the method has been tested on a battery-charging application using a DC-to-DC converter, it can also be extended in grid-connected applications by appropriate modification of the DC-to-alternating current (AC) inverter control. The system is built around a high-efficiency DC-to-DC converter and a low-cost microcontroller unit, which can easily perform additional operations such as battery-charging management or control of additional renewable energy sources (RES).

5.7 WG Total Efficiency Calculation

WG output power and losses are dependent on wind. To capture more energy from wind, it is important to analyze loss characteristics of the wind generator, which can be determined from wind speed. Furthermore, since many nonlinear losses occur in WGs, making prediction profit by using average wind speed may cause many errors. This section describes a method to represent various losses in the wind generator as a function of wind speed, which is based on the steady-state analysis. By using this method, wind turbine power, generated power, copper loss, iron loss, stray load loss, mechanical losses, and energy efficiency can be calculated quickly. However, this method cannot take a transient state into account. Generally, since the WG power change is large, a transient state seems to occur. To check the effect of this state the calculation is performed by using Power System Computer Aided Design (PSCAD)/Electromagnetic Transients in DC (EMTDC). It is seen that the difference between the two methods is negligible. By using wind speed data expressed by the probability density function, this method can predict energy produced by WGs and also estimate its total efficiency and capacity factor. As a whole, it can be summarized that the prediction method can contribute properly to the design as well as construction planning of a wind farm.

5.7.1 Outline of the Calculation Method

Induction generators are widely used as WGs due to their low cost, low maintenance, and direct grid connection. So an induction generator is considered for calculation efficiency. However, there are several problems regarding IGs:

- Usually, IG input, output, and loss conditions can be determined from rotational speed (slip). However, it is difficult to determine slip from wind turbine input torque.

Table 5.2 Wind Generators Losses

Mechanical loss	Gearbox losses
	Windage loss
	Ball-bearing loss
Copper loss	Primary winding copper loss
	Secondary winding copper loss
Iron loss	Eddy current loss
Stray load loss	Hysteresis loss

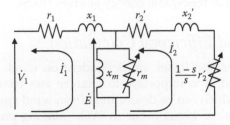

r_1 = stator resistance, r_2' = rotor resistance,
x_1 = stator leakage reactance, x_2' = rotor leakage
reactance, r_m = iron loss resistance,
x_m = magnetizing reactance, s(slip) = (Ns-N)/Ns,
N = rotor speed, Ns = synchronous speed.

Figure 5.8 Equivalent circuit of induction generator.

- Generator input torque is reduced by mechanical losses, but mechanical losses are a function of rotational speed (slip). It is difficult to determine mechanical losses and slip at the same time.
- It is hard to measure stray load loss and iron loss.
- It is difficult to evaluate gear loss analytically as a function of rotational speed.

This book describes a method of calculating the efficiency of WGs correctly, taking into account the aforementioned points. Table 5.2 shows the losses of wind generators. The equivalent circuit of the induction generator used in the proposed method is shown in Figure 5.8. The input torque and copper losses are calculated by solving the circuit equations (5.5).

$$\dot{V}_1 = -\left(r_1 + jx_1 + \frac{jr_m x_m}{r_m + jx_m} \right)\dot{I}_1 + \frac{jr_m x_m}{r_m + jx_m}\dot{I}_2$$

$$0 = -\frac{jr_m x_m}{r_m + jx_m}\dot{I}_1 + \left(\frac{jr_m x_m}{r_m + jx_m} + \frac{r_2'}{s} + jx_2 \right)\dot{I}_2$$

(5.5)

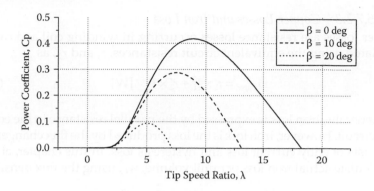

Figure 5.9 Power coefficient versus tip–speed ratio characteristics.

5.7.2   Equations for Analysis

5.7.2.1   Wind Turbine Power

The power captured from the wind can be expressed as (5.6). In this context, the MOD-2 model is used, whose power coefficient curve is shown in Figure 5.9. This turbine characteristic is nonlinear expressed as (5.8).

$$P_{wtb} = \frac{1}{2}\rho C_p(\lambda, \beta)\pi R^2 V_w^3 \quad [W] \tag{5.6}$$

$$\lambda = \frac{\omega_{wtb}R}{Vw} \tag{5.7}$$

$$C_P(\lambda, \beta) = 0.5(\lambda - 0.022\beta^2 - 5.6)e^{-0.17\lambda} \tag{5.8}$$

where P_{wtb} is the turbine output power (W), ρ is the air density (kg/m³), C_p is the power coefficient, λ is the tip–speed ratio, R is the radius of the blade (m), Vw is the wind speed (m/s), ω_{wtb} is the wind turbine angular speed (rad/s), and β is the blade pitch angle (degrees).

5.7.2.2   Generator Input

Generator input power can be calculated from the equivalent circuit of Figure 5.8:

$$I_2^2\left(\frac{1-s}{s}\right) \times r_2') \quad [W] \tag{5.9}$$

5.7.2.3   Copper Losses and Iron Loss

Copper losses are resistance losses occurring in winding coil and can be calculated using the equivalent circuit resistances, r_1 and r_2', as

$$w_{copper} = r_1 \times I_1^2 + r_2' \times I_2^2 \quad [W] \tag{5.10}$$

Generally, iron loss is expressed by the parallel resistance in the equivalent circuit. However, iron loss is the loss produced by the flux change, and it consists of eddy current loss and hysteresis loss. In this chapter, at first we calculate actual iron loss per unit volume, w_f, using the flux density:

$$w_f = B^2 \left\{ \sigma_H \left(\frac{f}{100} \right) + \sigma_E d^2 \left(\frac{f}{100} \right)^2 \right\} \quad [W/kg] \tag{5.11}$$

where B is the flux density (T), σ_H is the hysteresis loss coefficient, σ_E is the eddy current loss coefficient, f is the frequency (Hz), and d is the thickness of iron core steel plate (mm).

Normally, flux and internal voltage can be related to (5.12). Therefore, if the number of turns of a coil is fixed, a proportionality holds between the flux density and the internal voltage.

$$E = 4.44 \times f \times k_w \times w \times \phi \quad [V] \tag{5.12}$$

where k_w is the winding coefficient, w is the number of turns, and ϕ is the flux.

$$B = B_0 \times \frac{E}{E_0} \quad [T] \tag{5.13}$$

where E_0 is the nominal internal voltage.

And then the iron loss resistance can be obtained with respect to the internal voltage E determined by the flux density as shown in (5.14). W_f is the total iron loss, which is determined using (5.11) and the iron core weight.

$$r_m = \frac{E^2}{W_f/3} \tag{5.14}$$

5.7.2.4   Bearing Loss, Windage Loss, and Stray Load Loss

Bearing loss is a mechanical friction loss due to the rotation of the rotor:

$$W_b = K_B \omega_m \quad [W] \tag{5.15}$$

where K_B is a parameter concerning the rotor weight, the diameter of an axis, and the rotational speed of the axis.

Windage loss is a friction loss that occurs between the rotor and the air:

$$W_m = K_W \omega_m^2 \quad [W] \tag{5.16}$$

where K_w is a parameter determined by the rotor shape, its length, and the rotational speed.

Stray load loss is expressed as

$$W_s = 0.005 \frac{P^2}{P_n} \quad [W] \tag{5.17}$$

where P is generated power (W), and P_n is rated power (W).

5.7.2.5   Gearbox Loss [30]

Gearbox losses are primarily due to tooth contact losses and viscous oil losses. In general, these losses are difficult to predict. However, tooth contact losses are very small compared with viscous losses, and at fixed rotational speed viscous losses do not vary strongly with transmitted torque. Therefore, simple approximation of gearbox efficiency can be obtained by neglecting the tooth losses and assuming that the viscous losses are constant (a fixed percentage of the rated power). A viscous loss of 1% of rated power per step is a reasonable assumption. Thus, the efficiency of a gearbox with q steps can be computed using (5.18). Generally, the maximum gear ratio per step is approximately 6:1, so two or three steps of gears are typically required.

$$\eta_{gear} = \frac{P_t}{P_m} = \frac{P_m - (0.01)qP_{mR}}{P_m} \times 100 \ [\%] \tag{5.18}$$

where P_t is gear box output power, P_m is turbine power, and P_{mR} is rated turbine power.

Figure 5.10 shows the gearbox efficiency for three gear steps. In this chapter, three steps are assumed, according to a large-sized wind generator in recent years.

5.7.3   Calculation Method

The efficiency of a generator is determined using the previously described loss analysis method. The IG input, output, and loss conditions can be determined from rotational speed (slip). However, it is difficult to determine slip from wind turbine input torque. Therefore, we adopt an

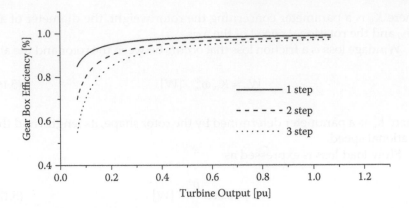

Figure 5.10 Gearbox efficiency.

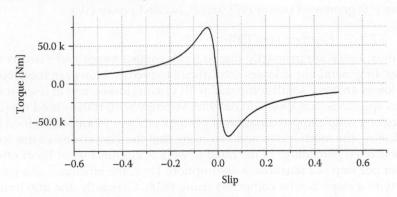

Figure 5.11 Slip–torque curve.

iterative calculation to obtain a slip, which produces torque equal to the wind turbine torque from a slip–torque curve as shown in Figure 5.11. Furthermore, it is difficult to determine mechanical losses and slip at the same time, because mechanical losses are a function of rotational speed (slip). Mechanical loss can also be obtained in the iterative calculation. The power transfer relation in the proposed method is shown in Figure 5.12.

Since mechanical losses and stray load loss cannot be expressed in a generator equivalent circuit, they have been deducted from the wind turbine output. Figure 5.13 shows the flow chart of the following proposed method:

1. Wind velocity is taken as the input value, and from this wind velocity all states of wind generators are calculated.
2. Wind turbine output is calculated from Equation (5.6). The synchronous angular velocity is taken as the initial value of the angular velocity, and wind turbine power is multiplied by the gear efficiency, η_{gear}.

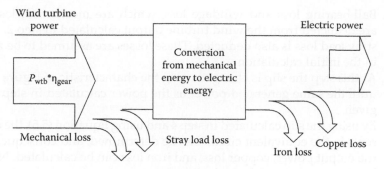

Figure 5.12 Expression of power flow in the proposed method.

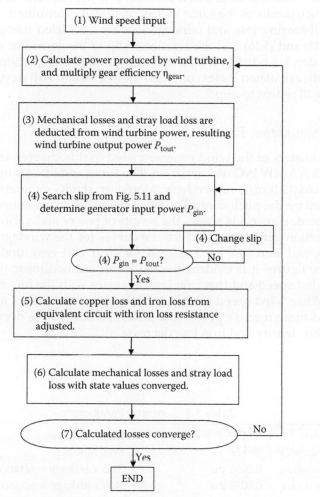

Figure 5.13 Flow chart of the proposed method.

3. Ball-bearing loss and windage loss, which are mechanical losses, are deducted from the wind turbine output calculated in step 2, and stray load loss is also deducted. These losses are assumed to be zero in the initial calculation.
4. At this step the slip is changed using the characteristic of Figure 5.11 until the same generated power as the power calculated in step 3 is given.
5. By using the slip calculated in step 4 and using Equation (5.5), the currents in the equivalent circuit can be determined, and consequently the output power, copper loss, and iron loss can be calculated. Next, loss W_f is calculated from the flux density using the previously outlined iron loss calculation method, and the iron loss resistance, r_m, which produces the same loss as W_f, is also determined.
6. Ball-bearing loss and windage loss are calculated using Equations (5.15) and (5.16) and the rotational slip of the generator determined in step 5. And stray load loss is calculated from Equation (5.17).
7. If the calculated losses converge, the calculation will stop; otherwise, it will return to step 2.

5.7.4   Simulation Results

The parameters of the wind generator used in this chapter are shown in Table 5.3. A 5 MW WG was assumed. The cut-in and rated wind velocities are 5.8 and 12.0 m/s, respectively. Moreover, the IG generated power is controlled by the pitch controller when the wind speed is over the rated wind speed. Figure 5.14 shows the results of power and various losses of the induction generator, in which the curves for the windage loss, bearing loss, and iron loss are enlarged for clear and easy understanding. From the figures it is evident that all losses are nonlinear with respect to the wind speed and that iron loss decreases with the increase of wind speed. When wind speed increases, IG real power increases, and thus the IG draws more reactive power and internal voltage of IG decreases. As a result, flux density and iron loss decrease.

Table 5.3 Generator Parameters

Rated power	5 MVA	Rated voltage	6,600 V
Rated frequency	60 Hz	Pole number	6
Stator resistance	0.0051 pu	Stator leakage reactance	0.088 pu
Rotor resistance	0.0091 pu	Rotor leakage reactance	0.125 pu
Iron resistance	1377.4 pu	Magnetizing reactance	4.776 pu

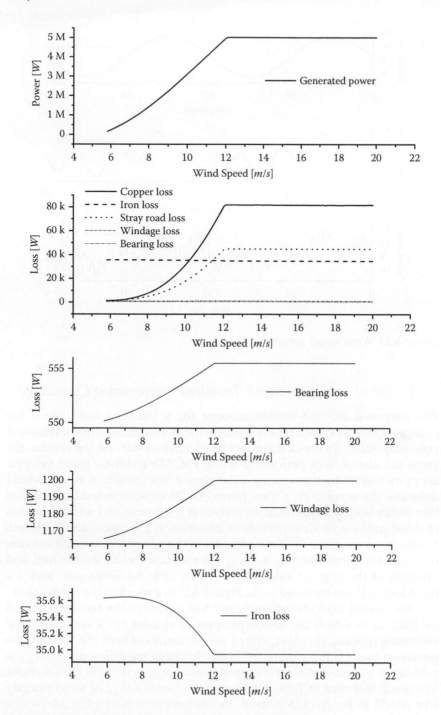

Figure 5.14 Power and various losses of induction generator.

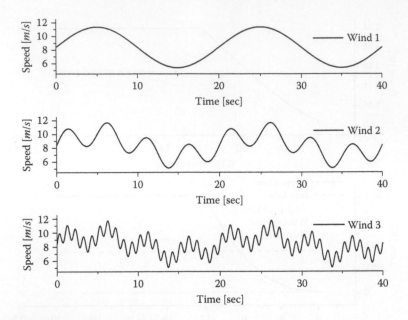

Figure 5.15 Wind speed variations.

5.7.5 Simulation Result with Transient Phenomenon Considered

The proposed method cannot account for a transient state. Since the change of the generated power of the wind generator is large, a transient state may occur. To check the effect of a transient state on the results, the same calculation was performed using PSCAD/EMTDC. Wind velocity is approximated by the sine waves as shown in Figure 5.15, where wind1 indicates the wave with a time period of 20 seconds, wind2 consists of two waves with periods of 20 seconds and 5 seconds, and wind3 consists of three waves with time periods of 20 seconds, 5 seconds, and 1 second. Moreover, to lengthen the time constant of the generator circuit, a calculation was also performed by setting the resistance value as one-half and one-fifth of the original value as shown in Table 5.4. Since gear loss is a fixed loss, it is not included here. Figure 5.16 is a result of the calculation.

The dotted lines shown in Figure 5.16 represent the results obtained by PSCAD in which the transient appears at time t = 1 sec due to the switching process. However, after 1 sec the results of both PSCAD and the proposed method are almost the same. To check the difference, energy is calculated by integrating the output power over one cycle (20 seconds). The result is shown in Table 5.4. Even if the large change of wind velocity like wind3 in Figure 5.15 is used, the difference is negligible. Moreover, even if the rotor resistance is small, the difference is not large. As a result,

Table 5.4 Calculation Results (I)

		Wind Turbine Energy (MJ)	Generated Power (MJ)	Energy Efficiency (%)
Wind1	PSCAD	43.88	42.40	96.64
	Proposed method	43.88	42.42	96.67
Wind2	PSCAD	42.04	40.07	96.81
	Proposed method	42.03	40.06	96.78
Wind3	PSCAD	40.96	39.70	96.9
	Proposed method	40.95	39.67	96.87
Wind3	PSCAD	40.96	39.80	97.14
$(r_2' \rightarrow 1/2)$	Proposed method	40.95	39.77	97.12
Wind3	PSCAD	40.96	39.85	97.29
$(r_2' \rightarrow 1/5)$	Proposed method	40.95	39.85	97.26

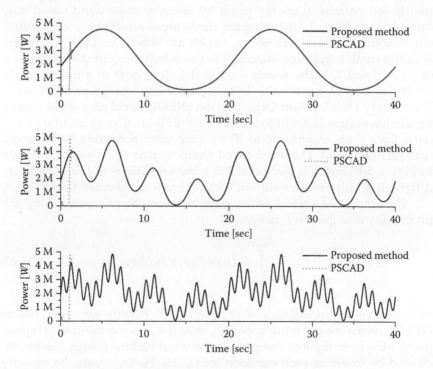

Figure 5.16 Output power of IG.

the difference between the two methods is so small that it can be ignored. Therefore, it can be said that the proposed method of efficiency calculation of the wind generator is sufficiently accurate.

5.7.6   Efficiency Calculation Using a Probability Density Function

If real wind information is available, the efficiency calculation of the wind generator can be precisely done. However, it is difficult to express real wind data as a function of time. Therefore, the Weibull function is used in this chapter. By using this function and the proposed method, it is possible to calculate the amount of annual power generation for a specific area. The Weibull function can be expressed by

$$f(v) = \frac{k}{c} \left(\frac{v}{c} \right)^{k-1} \exp\left[-\left(\frac{v}{c} \right)^k \right] \tag{5.19}$$

where k is shape factor, c is scale factor, and v is wind speed. From Equation (5.19), the status of the wind speed can be expressed. Although the annual average wind speed for two areas is the same, the probability functions for two areas can be different. Therefore, making prediction based on annual energy profit by using average wind speed may cause many errors. In this chapter, three areas are chosen, with different Weibull function parameters, which are shown in Table 5.5, where A is the weak wind area inland, B is the windy area in the tip portion of a cape, and C is the windy area at the slope part of a mountain. In Table 5.5, c and k were taken from Japan's New Energy and Industrial Technology Development Organization (NEDO) local area wind energy prediction system (LAWEPS) [31–32]. LAWEPS can give us accurate wind data. Data with an altitude of 70 m were used according to the wind turbine hub height. The wind speed characteristic of each area is shown in Figure 5.17, and it is clear that the wind conditions of these areas are different. In this case, the cut-out wind velocity is assumed to be 20 m/s.

The amount of annual power generation can be calculated using the probability distribution function:

$$E_{total} = \int_{V\min}^{V\max} P_g(v) \times f(v) \times 8670 dv \tag{5.20}$$

where E_{total} is the annual generated power (Wh), P_g is the generated power (W), V_{max} is the cut-out wind speed (m/s), and V_{min} is the cut-in wind speed (m/s). Moreover, the loss energy and the wind turbine energy can be calculated by inserting each equation into (5.16). Furthermore, the capacity factor and the total efficiency are calculated from

$$\text{Total Efficiency} = \frac{E_{total}[\text{Wh}]}{\text{Wind Turbine Energy}[\text{Wh}]} \times 100[\%] \tag{5.21}$$

Table 5.5 Weibull Parameters

	Area	Average Wind Speed (m/s)	c	k
A	Kyoto City	4.2	6.0	1.9
B	Erimo cape	9.0	10.0	2.2
C	Mt. Fuji's slope	10.5	9.5	1.6

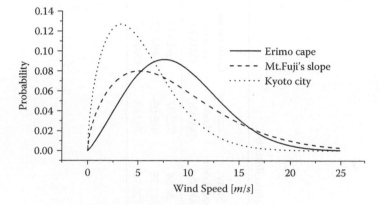

Figure 5.17 Probability density distribution.

$$\text{Capacity Factor} = \frac{E_{\text{total}}\left[\text{Wh}\right]}{\text{Rated Power}\left[\text{W}\right] \times 365 \times 24\left[\text{h}\right]} \times 100\left[\%\right] \quad (5.22)$$

Table 5.6 shows the simulation results. Though the average wind speed of area C is higher than that of area B, the capacity factor of area B is higher than that of area C. So the amount of power generation strongly depends on the probability density distribution. The difference of copper loss and stray load loss among the three areas is large because all areas depend not only on the generated power but also on the operating time, which is related to the probability density distribution. The differences of mechanical losses among the areas are not so large because they mainly depend on operating time. The influence of gearbox loss is comparatively greater in weak wind areas, such as in area A. Therefore, the total efficiency of area A is less than 90%.

5.8 Chapter Summary

This chapter describes the various types of wind energy conversion systems, namely, the fixed-speed and variable-speed wind generator

Table 5.6 Simulation Results

	Wind Turbine Energy (MWh)	Iron Loss (MWh)	Copper Loss (MWh)	Windage Loss (MWh)	Bearing Loss (MWh)	Stray Load Loss (MWh)	Gearbox Loss (MWh)	Generated Power (MWh)	Total Energy Efficiency (%)	Capacity Factor (%)
A	6669.42	123.76	58.61	4.10	1.92	34.66	522.37	5923.75	88.82	13.52
B	20696.07	227.38	252.39	7.63	3.56	141.78	965.05	19098.01	92.28	43.60
C	17198.42	186.69	214.09	6.26	2.92	120.11	792.65	15875.47	92.31	36.25

systems. Induction machines and synchronous machines are also discussed, as they are used mostly as wind generators. Relative advantages and disadvantages of various types of wind generators are discussed. The maximum power point tracking control system related to variable-speed wind generators is explained. Moreover, a method of WG total efficiency calculation is presented.

References

1. Y. Amirat, M. E. H. Benbouzid, B. Bensaker, R. Wamkeue, and H. Mangel, "The state of the art of generators for wind energy conversion systems," *Proceedings of the International Conference on Electrical Machine (ICEM'06)*, pp. 1–6, Chania, Greece, 2006.
2. E. Koutroulis and K. Kalaitzakis, "Design of a maximum power point tracking system for wind-energy-conversion-applications," *IEEE Transactions on Industrial Electronics*, vol. 53, no. 2, pp. 486–494, April 2006.
3. V. Valtchev, A. Bossche, J. Ghijselen, and J. Melkebeek, "Autonomous renewable energy conversion system," *Renewable Energy*, vol. 19, no. 1, pp. 259–275, January 2000.
4. E. Muljadi and C. P. Butterfield, "Pitch-controlled variable-speed wind turbine generation," *IEEE Trans. Ind. Appl.*, vol. 37, no. 1, pp. 240–246, January 2001.
5. L. H. Hansen et al., "Generators and power electronics technology for wind turbines," *Proceedings of IEEE IECON'01*, vol. 3, pp. 2000–2005, Denver, CO, November–December 2001.
6. T. Ackermann et al., "Wind energy technology and current status: A review," *Renewable and Sustainable Energy Reviews*, vol. 4, pp. 315–374, 2000.
7. R. W. Thresher et al., "Trends in the evolution of wind turbine generator configurations and systems," *Int. J. Wind Energy*, vol. 1, no. 1, pp. 70–86, April 1998.
8. P. Carlin et al., "The history and state of the art of variable-speed wind turbine technology," *Int. J. Wind Energy*, vol. 6, no. 2, pp. 129–159, April–June 2003.
9. A. Grauers et al., "Efficiency of three wind energy generator systems," *IEEE Trans. Energy Conversion*, vol. 11, no. 3, pp. 650–657, September 1996.
10. B. Blaabjerg et al., "Power electronics as efficient interface in dispersed power generation systems," *IEEE Trans. Power Electronics*, vol. 19, no. 5, pp. 1184–1194, September 2004.
11. P. Thoegersen et al., "Adjustable speed drives in the next decade. Future steps in industry and academia," *Electric Power Components & Systems*, vol. 32, no. 1, pp. 13–31, January 2004.
12. J. A. Baroudi et al., "A review of power converter topologies for wind generators," *Proceedings of IEEE IEMDC'05*, pp. 458–465, San Antonio, TX, May 2005.
13. B. Blaabjerg et al., "Power electronics as an enabling technology for renewable energy integration," *J. Power Electronics*, vol. 3, no. 2, pp. 81–89, April 2003.
14. C. Nicolas et al., "Guidelines for the design and control of electrical generator systems for new grid connected wind turbine generators," *Proceedings of IEEE IECON'02*, vol. 4, pp. 3317–3325, Seville, Spain, November 2002.

15. M. A. Khan et al., "On adapting a small pm wind generator for a multi-blade, high solidity wind turbine," *IEEE Trans. Energy Conversions*, vol. 20, no. 3, pp. 685–692, September 2005.

16. J. R. Bumby et al., "Axial-flux permanent-magnet air-cored generator for small-scale wind turbines," *IEE Proc. Electric Power Applications*, vol. 152, no. 5, pp. 1065–1075, September 2005.

17. G. K. Singh, "Self-excited induction generator research—A survey," *Electric Power Systems Research*, vol. 69, pp. 107–114, 2004.

18. R. C. Bansal et al., "Bibliography on the application of induction generators in nonconventional energy systems," *IEEE Trans. Energy Conversion*, vol. 18, no. 3, pp. 433–439, September 2003.

19. P. K. S. Khan et al., "Three-phase induction generators: A discussion on performance," *Electric Machines & Power Systems*, vol. 27, no. 8, pp 813–832, August 1999.

20. M. Ermis et al., "Various induction generator schemes for wind-electricity generation," *Electric Power Systems Research*, vol. 23, no. 1, pp. 71–83, 1992.

21. S. Muller et al., "Doubly fed induction generator systems for wind turbines e," *IEEE Industry Applications Magazine*, vol. 8, no. 3, pp. 26–33, May–June 2002.

22. R. Datta et al., "Variable-speed wind power generation using doubly fed wound rotor induction machine—A comparison with alternative scheme," *IEEE Trans. Energy Conversion*, vol. 17, no. 3, pp. 414–421, September 2002.

23. L. Holdsworth et al., "Comparison of fixed speed and doubly-fed induction wind turbines during power system disturbances," *IEE Proc. Generation, Transmission and Distribution*, vol. 150, no. 3, pp. 343–352, May 2003.

24. S. Grabic et al., "A comparison and trade-offs between induction generator control options for variable speed wind turbine applications," *Proceedings of IEEE ICIT'04*, vol. 1, pp. 564–568, Hammamet, Tunisia, December 2004.

25. P. Mutschler et al., "Comparison of wind turbines regarding their energy generation," *Proceedings of IEEE PESC'02*, vol. 1, pp. 6–11, Cairns, Australia, June 2002.

26. R. Hoffmann et al., "The influence of control strategies on the energy capture of wind turbines," *Proceedings of IEEE IAS'02*, vol. 2, pp. 886–893, Rome, Italy, October 2000.

27. M. Orabi et al., "Efficient performances of induction generator for wind energy," *Proceedings of IEEE IECON'04*, vol. 1, pp. 838–843, Busan, Korea, November 2004.

28. J. G. Slootweg et al., "Inside wind turbines—Fixed vs. variable speed," *Renewable Energy World*, pp. 30–40, 2003.

29. P. M. Anderson and A. Bose, "Stability simulation of wind turbine systems," *IEEE Trans. Power Apparatus and Systems*, vol. PAS-102, no. 12, pp. 3791–3795, December 1983.

30. G. L. Johnson, WIND ENERGY SYSTEMS Electronic Edition, http://www.rpc.com.au/products/windturbines/wind book/WindTOC.html

31. NEDO LAWEPS, http://www2.infoc.nedo.go.jp/nedo/top.html

32. NEDO, *The new energy and industrial technology development organization*, http://www.nedo.go.jp/english/introducing/what.html

33. M. Kayikci and J. V. Milanovic, "Assessing transient response of DFIG-based wind plants—The influence of model simplifications and parameters," *IEEE Trans. Power Syst.*, vol. 23, no. 2, pp. 545–554, May 2008.

34. D. J. Trudnowski, A. Gentile, J. M. Khan, and E. M. Petritz, "Fixed speed wind-generator and wind-park modeling for transient stability studies," *IEEE Trans. Power Syst.*, vol. 19, no. 4, pp. 1911–1917, November 2004.
35. B. Fox, D. Flynn, L. Bryans, N. Jenkins, D. Milborrow, M. O'Malley, et al., "Wind power integration: Connection and system operation aspects," *IET Power and Energy Series*, vol. 50, 2007.
36. T. Thiringer and J.-A. Dahlberg, "Periodic pulsations from a three bladed wind turbine," *IEEE Trans. Energy Conversion*, vol. 16, no. 2, pp. 128–133, June 2001.
37. D. S. L. Dolan and P. W. Lehn, "Simulation model of wind turbine 3P torque oscillations due to wind shear and tower shadow," *IEEE Trans. Energy Conversion*, vol. 21, no. 3, pp. 717–724, September 2006.
38. M. Kayikci and J. V. Milanovic, "Dynamic contribution of DFIG-based wind plants to system frequency disturbances," *IEEE Trans. Power Syst.*, vol. 24, no. 2, pp. 859–867, May 2009.
39. E. Muljadi and C. P. Butterfield, "Pitch-controlled variable-speed wind turbine generation," *IEEE Trans. Ind. Appl.*, vol. 37, no. 1, pp. 240–246, January–February 2001.
40. R. Cárdenas and R. Peña, "Sensorless vector control of induction machines for variable speed wind energy applications," *IEEE Trans. Energy Conversion*, vol. 19, no. 1, pp. 196–205, March 2004.
41. D. A. Torrey, "Switched reluctance generators and their control," *IEEE Transactions on Industrial Electronics*, vol. 49, no. 1, February 2002.
42. R. Cárdenas, "Control of a switched reluctance generator for variable-speed wind energy applications," *IEEE Transaction on Energy Conversion*, vol. 20, no. 4, December 2005.

chapter 6

Wind Generator Grid Integration Issues

6.1 Introduction

Wind power is often described as an "intermittent"—and therefore unreliable—energy source. In fact, at the power system level wind energy does not start and stop at irregular intervals, so the term *intermittent* is misleading. The output of aggregated wind capacity is variable, just as the power system itself is inherently variable. Since wind power production is dependent on the wind, the output of a turbine and wind farm varies over time under the influence of meteorological fluctuations. These variations occur on all time scales: by seconds, minutes, hours, days, months, seasons, and years. Understanding and predicting these variations is essential for successfully integrating wind power into the power system and to use it most efficiently. Wind power as a generation source has specific characteristics, which include variability and geographical distribution. These raise challenges for the integration of large amounts of wind power into electricity grids. To integrate large amounts of wind power successfully, a number of issues need to be addressed, including design and operation of the power system, grid infrastructure issues, and grid connection of wind power. This chapter discusses grid integration issues in detail.

6.2 Transient Stability and Power Quality Problems

Let us consider a grid-connected wind generator system. During a transient fault in the power network, the rotor speed of the wind generator goes very high, active power output goes very low, and terminal voltage goes very low or collapses. The wind speed might be considered constant during a transient fault.

According to grid code requirements, the voltage level should not be below 85% of the rated voltage. Usually the wind generator is shut down during these emergency situations. Recent tradition is not to shut down the wind generator during a network fault but to keep it connected to the grid through appropriate power electronics control. In other words, the wind generators should have fault ride-through (FRT) capability. This clearly indicates that wind generator stabilization is necessary during network faults.

6.3   Variability of Wind Power

Electricity flows—both supply and demand—are inherently variable because power systems are influenced by a large number of planned and unplanned factors, but they have been designed to cope effectively with these variations through their configuration, control systems, and interconnection [1].

Examples of demand include changing weather, which makes people switch their heating, cooling, and lighting on and off, and the instant power that millions of consumers expect for TVs and computers every day. On the supply side, when a large power station, especially a nuclear reactor, goes offline whether by accident or planned shutdown, it does so instantaneously, causing an immediate loss of many hundreds of megawatts. By contrast, wind energy does not suddenly trip the system off. Variations are smoother because there are hundreds or thousands of units rather than a few large power stations, making it easier for the system operator to predict and manage changes in supply. Especially in large, interconnected grids, there is little overall impact if the wind stops blowing in one particular place.

Predictability is key in managing wind power's variability, and significant advances have been made to improve forecasting methods. Today, wind power prediction is quite accurate for aggregated wind farms. Using increasingly sophisticated weather forecasts, wind power generation models, and statistical analysis, it is possible to predict generation from 5-minute to hourly intervals over time scales up to 72 hours in advance and for seasonal and annual periods. Using current tools, the forecast error for a single wind farm is between 10 and 20% of the power output for a forecast horizon of 36 hours. For regionally aggregated wind farms the forecast error is on the order of 10% for a day ahead and less than 5% for 1–4 hours in advance.

6.4   Power, Frequency, and Voltage Fluctuations Due to Random Wind Speed Variation

Due to random wind speed variation, wind generator output power, frequency, and terminal voltage fluctuate. In other words, power quality of the wind generator deteriorates. However, consumers need constant voltage and frequency. Thus, frequency, grid voltage, and transmission line power should be maintained constant. To this end, some control means are necessary [2–4].

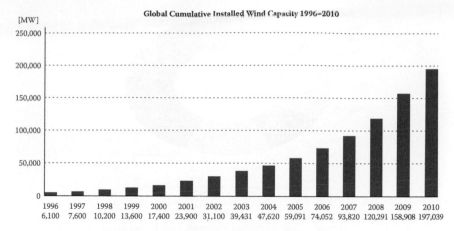

[MW]	1996	1997	1998	1999	2000	2001	2002	2003	2004	2005	2006	2007	2008	2009	2010
	6,100	7,600	10,200	13,600	17,400	23,900	31,100	39,431	47,620	59,091	74,052	93,820	120,291	158,908	197,039

Figure 1.1 Cumulative installed capacity, 1996–2010.

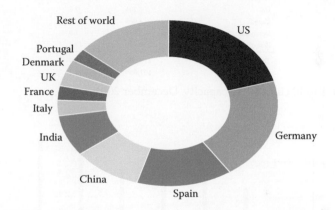

	MW	%
USA	25,170	20.8
Germany	23,903	19.8
Spain	16,754	13.9
China	12,210	10.1
India	9,615	8.0
Italy	3,736	3.1
France	3,404	2.8
UK	3,241	2.7
Denmark	3,180	2.6
Portugal	2,862	2.4
Rest of world	16,693	13.8
Total top 10	**104,104**	**85.2**
World total	**120,798**	**100.0**

Figure 1.2 Total installed capacity 2008.

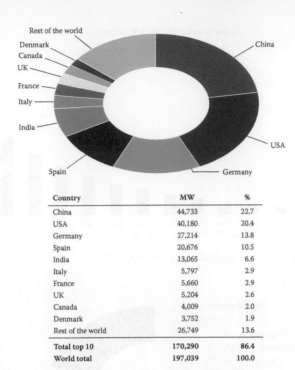

Country	MW	%
China	44,733	22.7
USA	40,180	20.4
Germany	27,214	13.8
Spain	20,676	10.5
India	13,065	6.6
Italy	5,797	2.9
France	5,660	2.9
UK	5,204	2.6
Canada	4,009	2.0
Denmark	3,752	1.9
Rest of the world	26,749	13.6
Total top 10	170,290	86.4
World total	197,039	100.0

Figure 1.3 Top 10 cumulative capacity, December 2010.

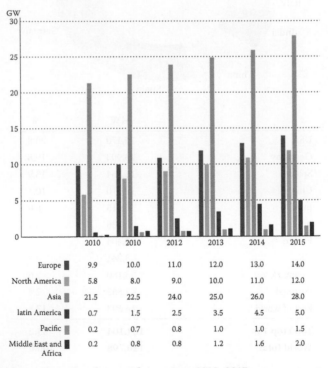

	2010	2010	2012	2013	2014	2015
Europe	9.9	10.0	11.0	12.0	13.0	14.0
North America	5.8	8.0	9.0	10.0	11.0	12.0
Asia	21.5	22.5	24.0	25.0	26.0	28.0
latin America	0.7	1.5	2.5	3.5	4.5	5.0
Pacific	0.2	0.7	0.8	1.0	1.0	1.5
Middle East and Africa	0.2	0.8	0.8	1.2	1.6	2.0

Figure 1.4 Annual market forecast by region, 2010–2015.

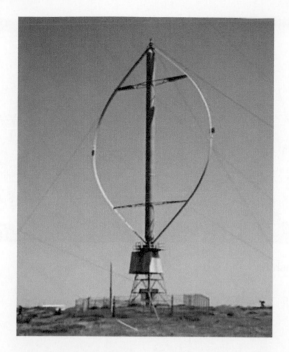

Figure 2.3 Vertical axis wind turbine.

Figure 2.4 Horizontal axis wind turbine.

Figure 2.5 Onshore wind farm.

Figure 2.6 Offshore wind turbine.

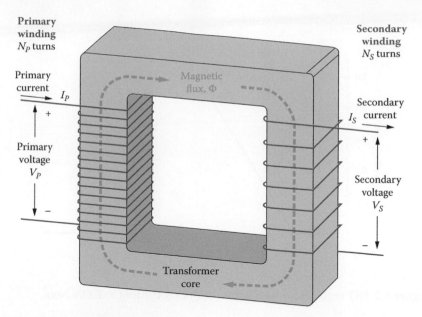

Figure 3.11 An ideal transformer.

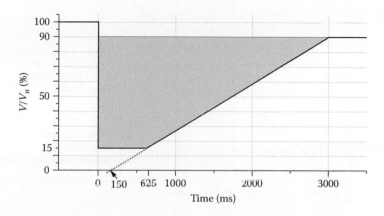

Figure 6.1 Typical low-voltage ride-through requirement—United States.

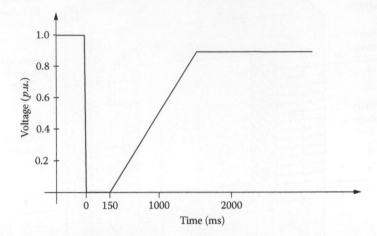

Figure 6.2 FRT requirements for wind turbines according to E.ON Netz.

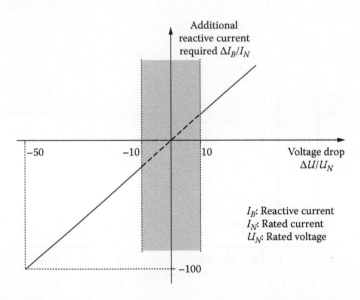

Figure 6.3 FRT requirements for wind turbines according to E.ON Netz.

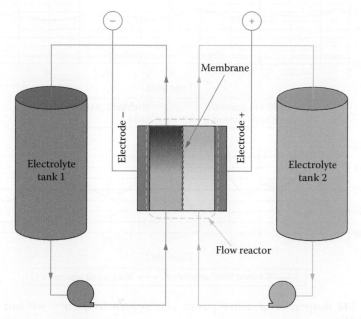

Figure 7.5 Flow battery cell.

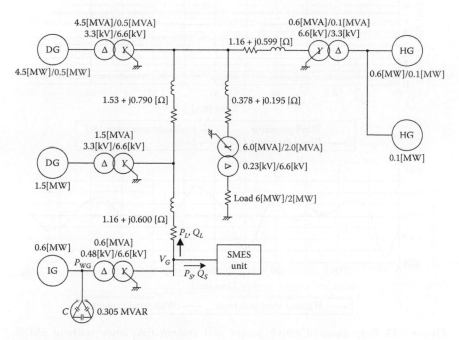

Figure 7.6 Power system model.

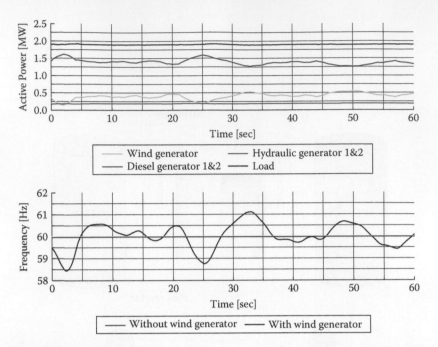

Figure 7.14 Responses of active power and system frequency without SMES (Load 2 MW).

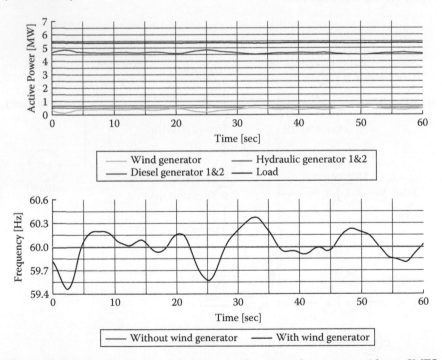

Figure 7.15 Responses of active power and system frequency without SMES (Load 6 MW).

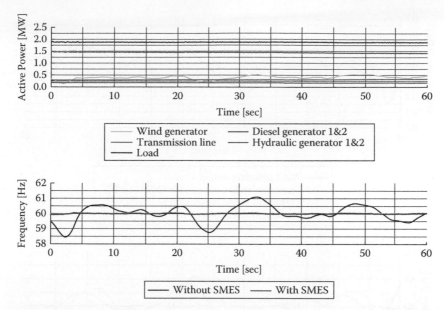

Figure 7.16 Responses of active power and system frequency with 10 MJ SMES (Load 2 MW).

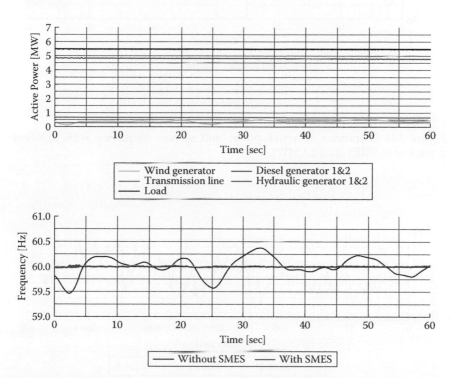

Figure 7.17 Responses of active power and system frequency with 10 MJ SMES (Load 6 MW).

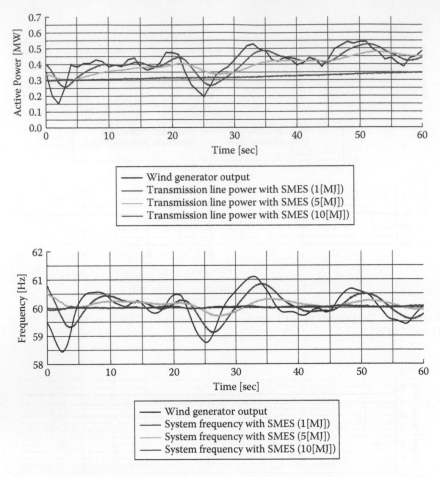

Figure 7.18 Responses of active power and system frequency with different capacities of SMES (Load 2 MW).

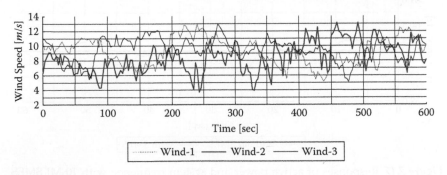

Figure 7.23 Wind speed.

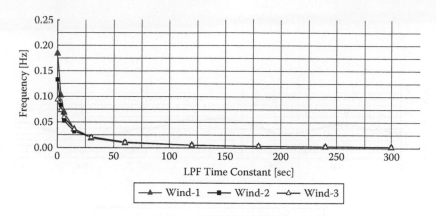

Figure 7.24 Maximum frequency fluctuation (WG capacity 10%).

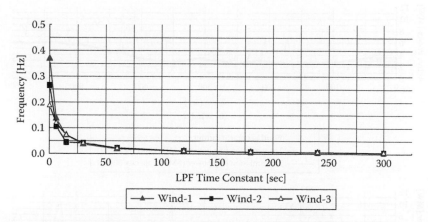

Figure 7.25 Maximum frequency fluctuation (WG capacity 20%).

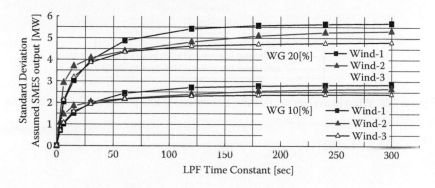

Figure 7.26 Assumed SMES output standard deviation.

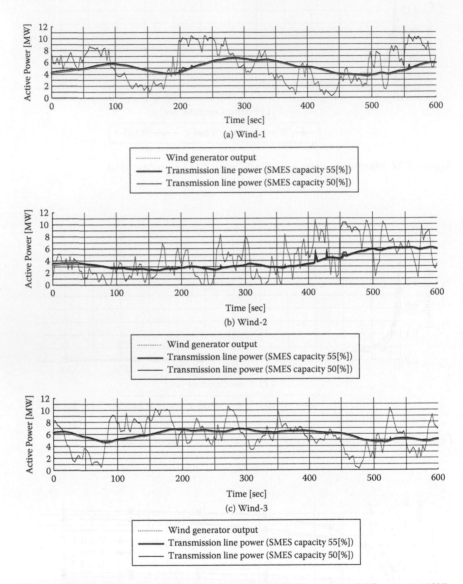

Figure 7.29 Wind generator output and transmission line power (WG capacity 10%).

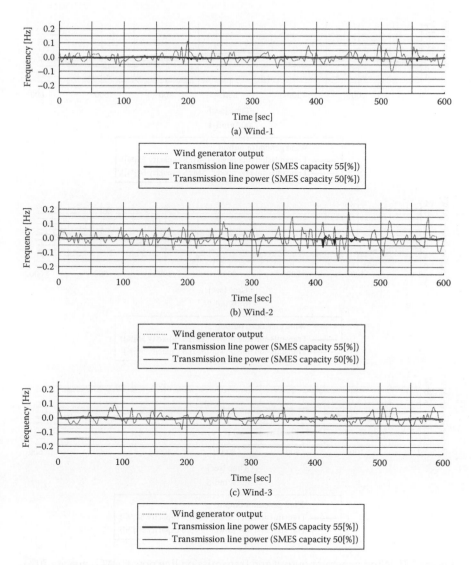

Figure 7.30 Frequency fluctuation (WG capacity 10%).

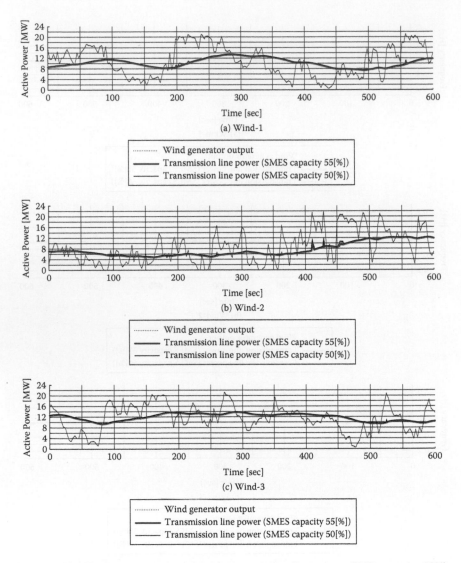

Figure 7.31 Wind generator output and transmission line power (WG capacity 20%).

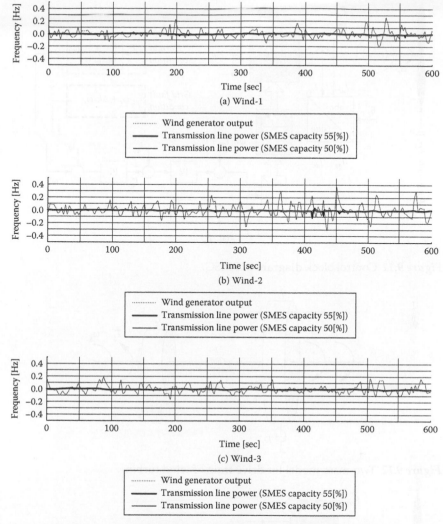

Figure 7.32 Frequency fluctuation (WG capacity 20%).

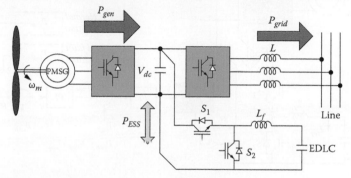

Figure 9.10 PMSG wind turbine system with ESS.

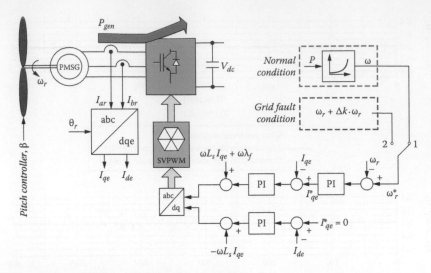

Figure 9.11 Control block diagram of PMSG.

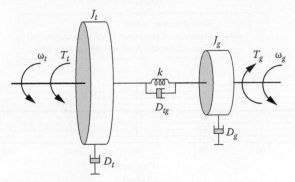

Figure 9.12 Two-mass model for drive train of wind turbine.

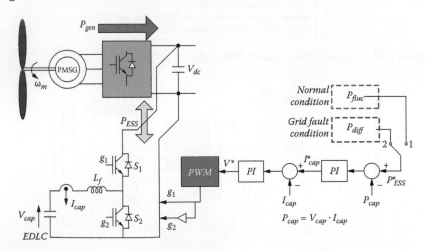

Figure 9.13 Control block diagram for ESS.

6.5 Grid Connection Requirements

With the rapid increase in installation of wind generators in the power system, it becomes necessary to require wind farms to behave as much as possible like conventional power plants to support the network voltage and frequency not only during steady-state conditions but also during grid disturbances. Due to this requirement, the utilities in many countries have recently established or are developing grid codes for operation and grid connection of wind farms. A grid code covers all material technical aspects relating to connections to, and the operation and use of, a country's electricity transmission system. They lay down rules that define the ways generating stations connecting to the system must operate to maintain grid stability. The aim of these grid codes is to ensure that the continued growth of wind generation does not compromise the power quality as well as the security and reliability of the electric power system [1].

Technical requirements within grid codes vary from system to system, but the typical requirements for generators normally concern tolerance, control of active and reactive power, protective devices, and power quality. Specific requirements for wind power generation are changing as penetration increases and as wind power is assuming more and more power plant capabilities, that is, assuming active control and delivering grid support services.

Wind turbine manufacturers have to respond to these grid code requirements. Therefore, much research has been conducted to develop technologies and solutions to meet these requirements. This section discusses major grid code requirements for operation and grid connection of wind farms. The major requirements of typical grid codes for operation and grid connection of wind turbines are summarized as follows.

1. Voltage operating range: The wind turbines are required to operate within typical grid voltage variations.
2. Frequency operating range: The wind turbines are required to operate within typical grid frequency variations.
3. Active power control: Several grid codes require wind farms to provide active power control to ensure a stable frequency in the system and to prevent overloading of lines. Also, wind turbines are required to respond with a ramp rate in the desired range.
4. Frequency control: Several grid codes require wind farms to provide frequency regulation capability to help maintain the desired network frequency.
5. Voltage control: Grid codes require that individual wind turbines control their own terminal voltage to a constant value by means of an automatic voltage regulator.

6. Reactive power control: The wind farms are required to provide dynamic reactive power control capability to maintain the reactive power balance and the power factor in the desired range.
7. Low-voltage ride-through (LVRT): In the event of a voltage sag, the wind turbines are required to remain connected for a specific amount of time before being allowed to disconnect. In addition, some utilities require that the wind turbines help support grid voltage during faults.
8. High-voltage ride-through (HVRT): In the event that the voltage goes above its upper limit value, the wind turbines should be capable of staying online for a given length of time.
9. Power quality: Wind farms are required to provide the electric power with a desired quality, such as maintaining constant voltage or voltage fluctuations in the desired range or maintaining voltage–current harmonics in the desired range.
10. Wind farm modeling and verification: Some grid codes require wind farm owners and developers to provide models and system data to enable the system operator to investigate by simulations the interaction between the wind farm and the power system. They also require installation of monitoring equipment to verify the actual behavior of the wind farm during faults and to check the model.
11. Communications and external control: The wind farm operators are required to provide signals corresponding to a number of parameters important for the system operator to enable proper operation of the power system. Moreover, it must be possible to connect and disconnect the wind turbines remotely.

Grid code requirements and regulations vary considerably from country to country and from system to system. The grid codes in a certain region or country may cover only a part of these requirements. The differences in requirements, besides local traditional practices, are caused by different wind power penetrations and by different degrees of power network robustness. Figure 6.1 shows typical LVRT requirements in the United States. Interconnection requirements for wind energy connected to the transmission networks in the United States are applicable to wind farms larger than 20 MW and mainly cover the following three major technical topics:

1. LVRT capability: The wind farms are required to remain online during voltage disturbances up to specified time periods and associated voltage levels, as described by Figure 6.1. According to this LVRT specification, the wind turbines should remain connected to the grid and supply reactive power when the voltage at the point of connection falls in the gray area. In addition, wind farms must be able to operate continuously at 90% of the rated line voltage, measured at the high-voltage side of the wind plant substation transformers.

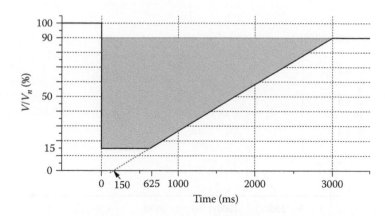

Figure 6.1 Typical low-voltage ride-through requirement—United States.

2. Power factor (reactive power) design criteria: The wind farms are required to maintain a power factor within the range of 0.95 leading to 0.95 lagging, measured at the high-voltage side of the substation transformers.
3. Supervisory control and data acquisition (SCADA) capability: The wind farms are required to have SCADA capability to transmit data and receive instructions from the transmission provider.

In response to increasing demands from the network operators—for example, to stay connected to the system during a fault event—the most recent wind turbine designs have been substantially improved. The majority of MW-size turbines being installed today are capable of meeting the most severe grid code requirements, with advanced features including fault ride-through capability. This enables them to assist in keeping the power system stable when disruptions occur. Modern wind farms are moving toward becoming wind energy power plants that can be actively controlled.

To illustrate how demanding the FRT requirements for wind turbines can be, the case of the German transmission system operator (TSO) E.ON Netz is presented. Figure 6.2 shows the requirement for the FRT process for wind turbines in the case of disturbance. According to the German code, wind turbines connected to the transmission system (which includes voltages levels of 220 kV and greater) must remain connected to the grid during short-circuit faults as long as the voltage at the point of common coupling (PCC), measured at the high-voltage level of the grid connected transformer, is above the continuous line defined in Figure 6.2. As can be observed, the turbines have to sustain the operation within the 150 ms after the fault appearance even if the voltage at the PCC decreases to zero [39].

In addition to the FRT capability, requirements regarding to voltage stability support are also imposed. Wind turbines must be able to provide

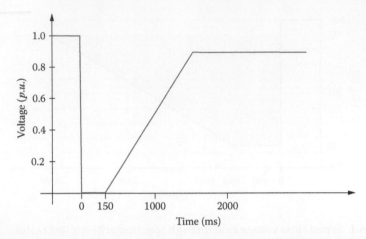

Figure 6.2 (See color insert.) FRT requirements for wind turbines according to E.ON Netz.

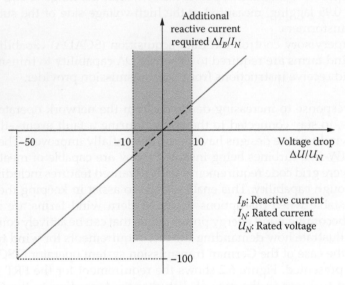

Figure 6.3 (See color insert.) FRT requirements for wind turbines according to E.ON Netz.

at least 2% of the rated reactive current for each percent of voltage dip as shown in Figure 6.3. A reactive current of 100% of the rated current must be possible if necessary. The reactive current must be injected into the network within 20 ms after the fault detection on the low-voltage side of the connecting transformer.

A dead band is also included around the reference voltage, in which the wind turbines can operate with power factor control. The requirements imposed by E.ON Netz in case of disturbance are technology oriented; that is, they address wind turbines manufacturer design requirements without explicitly defining power system security constraints or limits. Such required levels are well justified in power systems with high penetration levels of wind power (e.g., the northern area of Germany covered by E.ON Netz, with one of the highest penetration levels of wind power in the world) but are not necessary in power systems in an initial stage of wind power development. Requirements based on system security margins are an alternative to such grid codes.

Grid integration concerns have come to the fore in recent years as wind power penetration levels have increased in a number of countries as an issue that may impede the widespread deployment of wind power systems. Two of the strongest challenges to wind power's future prospects are the problems of intermittency and grid reliability.

Electricity systems must supply power in close balance to demand. The average load varies in predictable daily and seasonal patterns, but there is an unpredictable component due to random load variations and unforeseen events. To compensate for these variations, additional generation capacity is needed to provide regulation or to set aside as reserves. Generators within an electrical system have varying operating characteristics: Some are base-load plants; others, such as hydro- or combustion turbines, are more agile in terms of response to fluctuations and startup times. There is an economic value above the energy produced to a generator that can provide these ancillary services. Introducing wind generation can increase the regulation burden and need for reserves, due to its natural intermittency. The impact of the wind plant variability may range from negligible to significant depending on the level of penetration and intermittency of the wind resource.

6.5.1 Islanding and Auto Reclosure

Critical situations can occur if a part of the utility network is islanded, and an integrated distributed generation (DG) unit is connected. This situation is commonly referred to as loss of mains (LOM) or loss of grid (LOG). When LOM occurs, neither the voltage nor the frequency is controlled by the utility supply. Normally, islanding is the consequence of a fault in the network. If an embedded generator continues its operation after the utility supply was disconnected, faults may not clear since the arc is still charged. Small embedded generators (or grid interfaces, respectively) are often not equipped with voltage control; therefore, the voltage magnitude of an islanded network is not kept between desired limits, and undefined voltage magnitudes may occur during island operation.

Another result of missing control might be frequency instability. Since real systems are never balanced exactly, the frequency will change due to active power unbalance. Uncontrolled frequency represents a high risk for machines and drives. Since arc faults normally clear after a short interruption of the supply, automatic (instantaneous) reclosure is a common relay feature. With a continuously operating generator in the network, two problems may arise when the utility network is automatically reconnected after a short interruption:

- The fault may not have cleared since the arc was fed from the DG unit; therefore, instantaneous reclosure may not succeed.
- In the islanded part of the grid, the frequency may have changed due to active power unbalance. Reclosing the switch would couple two asynchronously operating systems.

Extended dead time has to be regarded between the separation of the DG unit and the reconnection of the utility supply to make fault clearing possible. Common time settings of auto reclosure relays are between 100 and 1,000 ms. With DG in the network, the total time has to be prolonged. A recommendation is to maintain a reclosure interval of 1 sec or more for distribution feeders with embedded generators.

The only solution to this problem seems to be to disconnect the DG unit as soon as LOM occurred. Thus, it is necessary to detect islands quickly and reliably.

6.5.2   Other Issues

There are some other problems concerning the integration of DG besides those already mentioned. These issues are already known from experience with conventional power systems.

6.5.2.1   Ferroresonance

Ferroresonance can occur and damage customer equipment or transformers. For cable lines, where faults are normally permanent, fast-blowing fuses are used as overcurrent protection. Since the fuses in the three phases do not trigger simultaneously, it may happen that a transformer is connected via only two phases for a short time. Then, the capacitance of the cable is in series with the transformer inductance that could cause distorted or high voltages and currents due to resonance conditions.

6.5.2.2   Grounding

There are possible grounding problems due to multiple ground current paths. If a DG unit is connected via a grounded delta-wye transformer, earth faults on the utility line will cause ground currents in both directions—from the fault to the utility transformer as well as to the DG transformer. This is normally not considered in the distribution system ground fault coordination. The problem of loss of earth (LOE) for single-point grounded distribution systems is that whenever the utility earth connection is lost the whole system gets ungrounded.

6.6 Design and Operation of Power Systems

Power systems have always had to deal with sudden output variations from large power plants, and the procedures put in place can be applied to deal with variations in wind power production as well. The issue is therefore not one of variability in itself but how to predict and manage this variability and what tools can be used to improve efficiency.

Wind power as a generation source has specific characteristics including variability and geographical distribution. These raise challenges for the integration of large amounts of wind power into electricity grids. To integrate large amounts of wind power successfully, a number of issues need to be addressed, including design and operation of the power system, grid infrastructure issues, and grid connection of wind power.

Experience has shown that the established control methods and system reserves available for dealing with variable demand and supply are more than adequate for coping with the additional variability from wind energy up to penetration levels of around 20%, depending of the nature of the system in question. This 20% figure is merely indicative, and the reality will vary widely from system to system. The more flexible a power system is in terms of responding to variations on both the demand and the supply side, the easier it to integrate variable-generation sources such as wind energy. In practice, such flexible systems, which tend to have higher levels of hydropower and gas generation in their power mix, will find that significantly higher levels of wind power can be integrated without major system changes.

Within Europe, Denmark already gets 21% of its gross electricity demand from the wind, Spain almost 12%, Portugal 9%, Ireland 8%, and Germany 7%. Some regions achieve much higher penetrations. In the western half of Denmark, for example, more than 100% of demand is sometimes met by wind power. Grid operators in a number of European countries, including Spain and Portugal, have now introduced central control centers that can monitor and manage efficiently the entire national fleet of wind turbines.

The present levels of wind power connected to electricity systems already show that it is feasible to integrate the technology to a significant extent.

Another frequent misunderstanding concerning wind power relates to the amount of "backup" generation capacity required, as the inherent variability of wind power needs to be balanced in a system.

Wind power does indeed have an impact on the other generation plants in a given power system, the magnitude of which will depend on the power system size, generation mix, load variations, demand size management, and degree of grid interconnection. However, large power systems can take advantage of the natural diversity of variable sources. They have flexible mechanisms to follow the varying load and plant outages that cannot always be accurately predicted.

6.7   Storage Options

There is increasing interest in both large-scale storage implemented at transmission level and smaller-scale dedicated storage embedded in distribution networks. The range of storage technologies is potentially wide. For large-scale storage, pumped hydro-accumulation storage (PAC) is the most common and known technology and can also be done underground. Another technology option available on a large scale is compressed air energy storage (CAES).

On a decentralized scale, storage options include flywheels, batteries (possibly in combination with electric vehicles), fuel cells, electrolysis, and super-capacitors. Furthermore, an attractive solution consists of the installation of heat boilers at selected combined heat and power locations (CHPs) order to increase the operational flexibility of these units.

It should be pointed out that storage leads to energy losses and is not necessarily an efficient option for managing wind farm output. If a country does not have favorable geographical conditions for hydroreservoirs, storage is not an attractive solution because of the poor economics at moderate wind power penetration levels (up to 20%). In any case, the use of storage to balance variations at wind plant level is neither necessary nor economic.

6.8   Grid Infrastructure

The specific nature of wind power as a distributed and variable-generation source requires specific infrastructure investments and the implementation of new technology and grid management concepts. High levels of wind energy in a system can impact grid stability, congestion management, transmission efficiency, and transmission adequacy.

In many parts of the world, substantial upgrades of grid infrastructure will be required to allow for significant levels of grid integration. Great improvements can be achieved by network optimization and other "soft" measures, but an increase in transmission capacity and construction of new transmission lines will also be needed. At the same time, adequate and fair procedures for grid access for wind power need to be developed and implemented, even in areas where grid capacity is limited. However, the expansion of wind power is not the only driver. Extensions and reinforcements are needed to accommodate whichever power generation technology is chosen to meet a rapidly growing electricity demand.

6.9 Wind Power's Contribution to System Adequacy

The *capacity credit* of wind energy expresses how much "conventional" power generation capacity can be avoided or replaced by wind energy. For low wind energy penetration levels, the capacity credit will therefore be close to the average wind power production, which depends on the capacity factors on each individual site (normally 20–35% of rated capacity). With increasing penetration levels of wind power, its relative capacity credit will decrease, which means that a new wind plant on a system with high wind power penetration will replace less conventional power than the first plants in the system. Aggregated wind plants over larger geographical areas are best suited to take full advantage of the firm contribution of wind power in a power system.

6.10 Chapter Summary

This chapter provides a thorough description of the grid integration issues of wind generator systems. Wind power as a generation source has specific characteristics, which include variability and geographical distribution. These raise challenges for the integration of large amounts of wind power into electricity grids. To integrate large amounts of wind power successfully, a number of issues need to be addressed, including design and operation of the power system, grid infrastructure issues, and grid connection of wind power, power quality, and transient stability enhancement of wind generator systems. This chapter focuses on all of these phenomena.

References

1. Global Wind Energy Council (GWEC), http://www.gwec.net/
2. M. H. Ali and B. Wu, "Comparison of stabilization methods for fixed-speed wind generator systems," *IEEE Transactions on Power Delivery*, vol. 25, no. 1, pp. 323–331, January 2010.
3. M. H. Ali, J. Tamura, and B. Wu, "SMES strategy to minimize frequency fluctuations of wind generator system," *Proceedings of the 34th Annual Conference of the IEEE Industrial Electronics Society (IECON 2008)*, November 10–13, 2008, Orlando, FL, pp. 3382–3387.
4. M. H. Ali, T. Murata, and J. Tamura, "Minimization of fluctuations of line power and terminal voltage of wind generator by fuzzy logic-controlled SMES," *International Review of Electrical Engineering (IREE)*, vol. 1, no. 4, pp. 559–566, October 2006.

5. M. Abbes, J. Belhadj, and A. B. Bennani, "Design and control of a direct drive wind turbine equipped with multilevel converters," *Renewable Energy*, vol. 35, pp. 936–945, 2010.

6. O. Abdel-Baqi and A. Nasiri, "A dynamic LVRT solution for doubly fed induction generators," *IEEE Transactions on Power Electronics*, vol. 25, pp. 193–196, 2010.

7. Y. M. Atwa, E. F. El-Saadany, and A. C. Guise, "Supply adequacy assessment of distribution system including wind based DG during different modes of operation," *IEEE Transactions on Power System*, vol. 25, pp. 78–86, 2010.

8. H. Banakar and B. T. Ooi, "Clustering of wind farms and its sizing impact," *IEEE Transactions on Energy Conversion*, vol. 24, pp. 935–942, 2009.

9. L. R. Chang-Chien and Y. C. Yin, "Strategies for operating wind power in a similar manner of conventional power plant," *IEEE Transactions on Energy Conversion*, vol. 24, pp. 926–934, 2009.

10. S. Z. Chen, N. C. Cheung, K. C. Wong, and J. Wu, "Grid synchronization of doubly-fed induction generator using integral variable structure control," *IEEE Transactions on Energy Conversion*, vol. 24, pp. 875–883, 2009.

11. R. Doherty, A. Mullane, G. Nolan, D. J. Burke, A. Bryson, and M. O'Malley, "An assessment of the impact of wind generation on system frequency control," *IEEE Transactions on Power Systems*, vol. 25, pp. 452–460, 2010.

12. J. B. Hu and Y. K. He, "Reinforced control and operation of DFIG based wind-power-generation system under unbalanced grid voltage conditions," *IEEE Transactions on Energy Conversion*, vol. 24, pp. 905–915, 2009.

13. M. V. Kazemi, A. S. Yazdankhah, and H. M. Kojabadi, "Direct power control of DFIG based on discrete space vector modulation," *Renewable Energy*, vol. 35, pp. 1033–1042, 2010.

14. A. Kusiak and Z. Song, "Design of wind farm layout for maximum wind energy capture," *Renewable Energy*, vol. 35, pp. 685–694, 2010.

15. A. Mendonca and J. A. P. Lopes, "Robust tuning of power system stabilizers to install in wind energy conversion systems," *IET Renewable Power Generation*, vol. 3, pp. 465–475, 2009.

16. Y. Mishra, S. Mishra, F. X. Li, Z. Y. Dong, and R. C. Bansal, "Small-signal stability analysis of a DFIG-based wind power system under different modes of operation," *IEEE Transactions on Energy Conversion*, vol. 24, pp. 972–982, 2009.

17. B. C. Ni and C. Sourkounis, "Influence of wind-energy-converter control methods on the output frequency components," *IEEE Transactions on Industry Applications*, vol. 45, pp. 2116–2122, 2009.

18. S. Nishikata and F. Tatsuta, "A new interconnecting method for wind turbine/generators in a wind farm and basic performances of the integrated system," *IEEE Transactions on Industrial Electronics*, vol. 57, pp. 468–475, 2010.

19. L. F. Ochoa, C. J. Dent, and G. P. Harrison, "Distribution network capacity assessment: Variable DG and active networks," *IEEE Transactions on Power Systems*, vol. 25, pp. 87–95, 2010.

20. G. Ramtharan, A. Arulampalam, J. B. Ekanayake, F. M. Hughes, and N. Jenkins, "Fault ride through of fully rated converter wind turbines with AC and DC transmission systems," *IET Renewable Power Generation*, vol. 3, pp. 426–438, 2009.

21. Y. Sozer and D. A. Torrey, "Modeling and control of utility interactive inverters," *IEEE Transactions on Power Electronics*, vol. 24, pp. 2475–2483, 2009.

22. M. Tazil, V. Kumar, R. C. Bansal, S. Kong, Z. Y. Dong, and W. Freitas, "Three-phase doubly fed induction generators: An overview," *IET Electric Power Applications*, vol. 4, pp. 75–89, 2010.

23. J. Usaola, "Probabilistic load flow in systems with wind generation," *IET Generation Transmission & Distribution*, vol. 3, pp. 1031–1041, 2009.

24. E. Vittal, M. O'Malley, and A. Keane, "A steady-state voltage stability analysis of power systems with high penetrations of wind," *IEEE Transactions on Power Systems*, vol. 25, pp. 433–442, 2010.

25. Y. Wang and L. Xu, "Coordinated control of DFIG and FSIG based wind farms under unbalanced grid conditions," *IEEE Transactions on Power Delivery*, vol. 25, pp. 367–377, 2010.

26. S. Zhang, K. J. Tseng, and S. S. Choi, "Statistical voltage quality assessment method for grids with wind power generation," *IET Renewable Power Generation*, vol. 4, pp. 43–54, 2010.

27. P. Zhou, Y. K. He, and D. Sun, "Improved direct power control of a DFIG-based wind turbine during network unbalance," *IEEE Transactions on Power Electronics*, vol. 24, pp. 2465–2474, 2009.

28. Y. M. Atwa, E. F. El-Saadany, M. M. A. Salama, and R. Seethapathy, "Optimal renewable resources mix for distribution system energy loss minimization," *IEEE Transactions on Power Systems*, vol. 25, pp. 360–370, 2010.

29. T. Ayhan and H. Al Madani, "Feasibilty study of renewable energy powered seawater desalination technology using natural vacuum technique," *Renewable Energy*, vol. 35, pp. 506–514, 2010.

30. J. L. Bernal-Agustin and R. Dufo-Lopez, "Techno-economical optimization of the production of hydrogen from PV-wind systems connected to the electrical grid," *Renewable Energy*, vol. 35, pp. 747–758, 2010.

31. H. C. Chiang, T. T. Ma, Y. H. Cheng, J. M. Chang, and W. N. Chang., "Design and implementation of a hybrid regenerative power system combining grid-tie and uninterruptible power supply functions," *IET Renewable Power Generation*, vol. 4, pp. 85–99, 2010.

32. P. D. Friedman, "Evaluating economic uncertainty of municipal wind turbine projects," *Renewable Energy*, vol. 35, pp. 484–489, 2010.

33. C. H. Liu, K. T. Chau, and X. D. Zhang, "An efficient wind photovoltaic hybrid generation system using doubly excited permanent-magnet brushless machine," *IEEE Transactions on Industrial Electronics*, vol. 57, pp. 831–839, 2010.

34. M. S. Lu, C. L. Chang, W. J. Lee, and L. Wang, "Combining the wind power generation system with energy storage equipment," *IEEE Transactions on Industry Applications*, vol. 45, pp. 2109–2115, 2009.

35. J. M. Morales, A. J. Conejo, and J. Perez-Ruiz, "Short-term trading for a wind power producer," *IEEE Transactions on Power Systems*, vol. 25, pp. 554–564, 2010.

36. R. Sebastian and R. P. Alzola, "Effective active power control of a high penetration wind diesel system with a Ni-Cd battery energy storage," *Renewable Energy*, vol. 35, pp. 952–965, 2010.

37. R. K. Varma, V. Khadkikar, and R. Seethapathy, "Nighttime application of PV solar farm as STATCOM to regulate grid voltage," *IEEE Transactions on Energy Conversion*, vol. 24, pp. 983–985, 2009.

38. D. L. Yao, S. S. Choi, K. J. Tseng, and T. T. Lie, "A statistical approach to the design of a dispatchable wind power-battery energy storage system," *IEEE Transactions on Energy Conversion*, vol. 24, pp 916–925, 2009.

39. C. Rahmann, H.-J. Haubrich, A. Moser, R. Palma-Behnke, L. Vargas, and M. B. C. Salles, "Justified Fault-Ride-Through Requirements for Wind Turbines in Power Systems," *IEEE Trans. Power Systems*, vol. 26, no. 30, pp. 1555–1563, August 2011.

chapter 7

Solutions for Power Quality Issues of Wind Generator Systems

7.1 Introduction

Using wind power to generate electricity is receiving more and more attention every day all over the world. One of the simplest methods of running a wind generation system is to use an induction generator (IG) connected directly to the power grid, because induction generators are the most cost-effective and robust machines for energy conversion. However, induction generators require reactive power for magnetization, particularly during startup. As the reactive power drained by the induction generators is coupled to the active power generated by them, the variability of wind speed results in variations of induction generators' real and reactive powers. It is this variation in active and reactive powers that interacts with the network and provokes voltage and frequency fluctuations. These fluctuations cause lamp flicker and inaccuracy in the timing devices. If good penetration of the wind power is to be achieved, some remedial measures must be taken for power quality improvement. Since both frequency and voltages are often affected in these systems, fast-acting control devices (i.e., the energy storage devices) capable of exchanging active as well as reactive powers are appropriate candidates to meet this end. This chapter discusses various energy storage devices, comparison among them, and the use of energy storage devices in minimizing fluctuations in line power, frequency, and terminal voltage of wind generator systems. Furthermore, this chapter discusses the output power leveling of wind generator systems by pitch angle control, power quality improvement by flywheel energy storage system, and constant power control of doubly fed induction generator (DFIG) wind turbines with supercapacitor energy storage.

7.2 Various Energy Storage Systems

Various promising energy storage systems are available on the market battery, such as energy storage, supercapacitor energy storage, superconducting magnetic energy storage (SMES), flywheel energy storage, and compressed air energy storage [1].

Electrical energy in an alternating current (AC) system cannot be stored electrically. However, energy can be stored by converting the AC electricity and storing it electromagnetically, electrochemically, or kinetically or as potential energy. Each energy storage technology usually includes a power conversion unit to change the energy from one form to another. Two factors characterize the application of an energy storage technology. One is the amount of energy that can be stored in the device. This is a characteristic of the storage device itself. Another is the rate at which energy can be transferred into or out of the storage device. This depends mainly on the peak power rating of the power conversion unit but is also impacted by the response rate of the storage device itself. The power–energy ranges for near-to-midterm technologies are projected in Figure 7.1. Integration of these four possible energy storage technologies with flexible AC transmission systems (FACTS) and custom power devices are among the possible power applications of energy storage. The possible benefits include transmission enhancement, power oscillation damping, dynamic voltage stability, tie line control, short-term spinning reserve, load leveling, underfrequency load shedding reduction, circuit-breaker reclosing, subsynchronous resonance damping, and power quality improvement.

7.2.1 Superconducting Magnetic Energy Storage

Although superconductivity was discovered in 1911, it was not until the 1970s that SMES was first proposed as an energy storage technology for power systems. SMES systems have attracted the attention of both electric

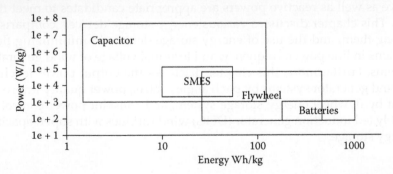

Figure 7.1 Specific power versus specific energy ranges for near-to-midterm technology.

utilities and the military due to their fast response and high efficiency; they have a charge–discharge efficiency over 95%. Possible applications include load leveling, dynamic stability, transient stability, voltage stability, frequency regulation, transmission capability enhancement, and power quality improvement [1].

When compared with other energy storage technologies, today's SMES systems are still costly. However, the integration of an SMES coil into existing FACTS devices eliminates what is typically the largest cost for the entire SMES system—the inverter unit. Some studies have shown that micro (0.1 MWh) and midsize (0.1–100 MWh) SMES systems could potentially be more economical for power transmission and distribution applications. The use of high-temperature superconductors should also make SMES cost-effective due to reductions in refrigeration needs. There are a number of ongoing SMES projects currently installed or in development throughout the world.

An SMES unit is a device that stores energy in the magnetic field generated by the direct current (DC) current flowing through a superconducting coil. The inductively stored energy (E in joules) and the rated power (P in watts) are commonly given specifications for SMES devices, and they can be expressed as follows:

$$E = \frac{1}{2}LI^2$$

$$P = \frac{dE}{dt} = LI\frac{dI}{dt} = VI$$

(7.1)

where L is the inductance of the coil, I is the DC current flowing through the coil, and V is the voltage across the coil. Since energy is stored as circulating current, energy can be drawn from an SMES unit with almost instantaneous response with energy stored or delivered over periods ranging from a fraction of a second to several hours.

An SMES unit consists of a large superconducting coil at the cryogenic temperature. This temperature is maintained by a cryostat or dewar that contains helium or nitrogen liquid vessels. A power conversion/conditioning system (PCS) connects the SMES unit to an AC power system, and it is used to charge and discharge the coil. Two types of power conversion systems are commonly used. One option uses a current source converter (CSC) to both interface to the AC system and charge and discharge the coil. The second option uses a voltage source converter (VSC) to interface to the AC system and a DC-to-DC chopper to charge and discharge the coil. The VSC and DC-to-DC chopper share a common DC bus. The components of an SMES system are shown in Figure 7.2. The modes of charge–discharge–standby are obtained by controlling the voltage across

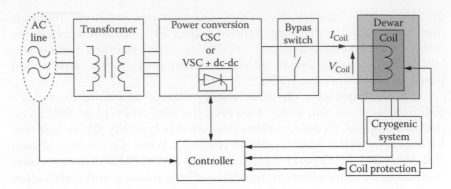

Figure 7.2 Components of a typical SMES system.

the SMES coil (V_{coil}). The SMES coil is charged or discharged by applying a positive or negative voltage, V_{coil}, across the superconducting coil. The SMES system enters a standby mode operation when the average V_{coil} is zero, resulting in a constant average coil current, I_{coil}.

Several factors are taken into account in the design of the coil to achieve the best possible performance of an SMES system at the least cost. These factors may include coil configuration, energy capability, structure, and operating temperature. A compromise is made between each factor by considering the parameters of energy/mass ratio, Lorentz forces, and stray magnetic field and by minimizing the losses for a reliable, stable, and economic SMES system. The coil can be configured as a solenoid or a toroid. The solenoid type has been used widely due to its simplicity and cost-effectiveness, though the toroid coil designs were also incorporated by a number of small-scale SMES projects. Coil inductance (L) or PCS maximum voltage (V_{max}) and current (I_{max}) ratings determine the maximum energy/power that can be drawn or injected by an SMES coil. The ratings of these parameters depend on the application type of SMES. The operating temperature used for a superconducting device is a compromise between cost and the operational requirements. Low-temperature superconductor devices (LTSs) are available now, whereas high-temperature superconductor devices are currently in the development stage. The efficiency and fast response capability (milliwatts/millisecond) of SMES systems have been and can be further exploited in applications at all levels of electric power systems. The potential utility applications have been studied since the 1970s. SMES systems have been considered for the following: (1) load leveling; (2) frequency support (spinning reserve) during loss of generation; (3) enhancing transient and dynamic stability; (4) dynamic voltage support (VAR compensation); (5) improving power quality; and (6) increasing transmission line capacity, thus enhancing overall reliability of power systems. Further development continues in power conversion systems and

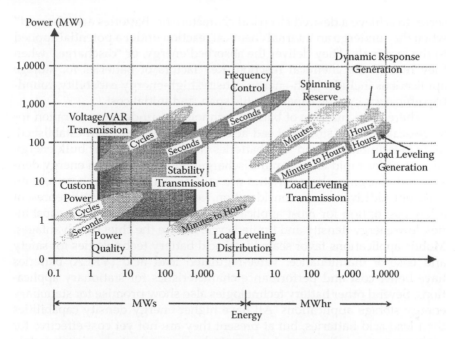

Figure 7.3 Energy–power characteristics of potential SMES applications.

control schemes, evaluation of design and cost factors, and analyses for various SMES system applications. The energy and power characteristics for potential SMES applications for generation, transmission, and distribution are depicted in Figure 7.3. The square area in the figure represents the applications that are currently economical. Therefore, the SMES technology has a unique advantage in two types of application: power system transmission control; and stabilization and power quality.

The cost of an SMES system can be separated into two independent components: (1) the cost of the energy storage capacity; and (2) the cost of the power handling capability. Storage-related cost includes the capital and construction costs of the conductor, coil structure components, cryogenic vessel, refrigeration, protection, and control equipment. Power-related cost involves the capital and construction costs of the power conditioning system. While the power-related cost is lower than the energy-related cost for large-scale applications, it is more dominant for small-scale applications.

7.2.2 Battery Energy Storage Systems

Batteries are one of the most cost-effective energy storage technologies available, with energy stored electrochemically. A battery system is made up of a set of low-voltage/power battery modules connected in parallel and series to achieve a desired electrical characteristic. Batteries are "charged" when they undergo an internal chemical reaction under a potential applied to the terminals. They deliver the absorbed energy, or "discharge," when they reverse the chemical reaction. Key factors of batteries for storage applications include: high-energy density, high-energy capability, round-trip efficiency, cycling capability, life span, and initial cost.

There are a number of battery technologies under consideration for large-scale energy storage. Lead acid batteries represent an established, mature technology. Lead acid batteries can be designed for bulk energy storage or for rapid charge and discharge. Improvements in energy density and charging characteristics are still an active research area, with different additives under consideration. Lead acid batteries still represent a low-cost option for most applications requiring large storage capabilities; low-energy density and limited cycle life are the chief disadvantages. Mobile applications favor sealed lead acid battery technologies for safety and ease of maintenance. Valve-regulated lead acid (VRLA) batteries have better cost and performance characteristics for stationary applications. Several other battery technologies also show promise for stationary energy storage applications. All have higher energy density capabilities than lead acid batteries, but at present they are not yet cost-effective for higher-power applications. Leading technologies include nickel–metal–hydride batteries, nickel–cadmium batteries, and lithium–ion batteries. The last two technologies are both being pushed for electric vehicle applications where high-energy density can offset higher cost to some degree.

Due to the chemical kinetics involved, batteries cannot operate at high power levels for long time periods. In addition, rapid, deep discharges may lead to early replacement of the battery, since heating resulting in this kind of operation reduces battery lifetime. There are also environmental concerns related to battery storage due to toxic gas generation during battery charge and discharge. The disposal of hazardous materials presents some battery disposal problems. The disposal problem varies with battery technology. For example, the recycling and disposal of lead acid batteries is well established for automotive batteries. Batteries store DC charge, so power conversion is required to interface a battery with an AC system. Small, modular batteries with power electronic converters can provide four-quadrant operation (bidirectional current flow and bidirectional voltage polarity) with rapid response. Advances in battery technologies offer increased energy storage densities, greater cycling capabilities, higher reliability, and lower cost. Battery energy storage systems (BESSs) have recently emerged as one of the more promising near-term storage technologies for power applications, offering a wide range of power system applications such as area regulation, area protection, spinning reserve, and power factor correction. Several BESS units have been designed and installed in existing systems for the purposes of load leveling, stabilizing, and load frequency control. Optimal installation site and capacity of BESSs can be determined depending on their application. This has been done for load-leveling applications. Also, the integration of battery energy storage with a FACTS power flow controller can improve the power system operation and control.

7.2.3 Advanced Capacitors

Capacitors store electric energy by accumulating positive and negative charges (often on parallel plates) separated by an insulating dielectric. The capacitance, C, represents the relationship between the stored charge, q, and the voltage between the plates, V, as shown in (7.2). The capacitance depends on the permittivity of the dielectric, e, the area of the plates, A, and the distance between the plates, d, as shown in (7.3). Equation (7.4) shows that the energy stored on the capacitor depends on the capacitance and on the square of the voltage.

$$q = CV \tag{7.2}$$

$$C = \frac{\varepsilon A}{d} \tag{7.3}$$

$$E = \frac{1}{2}CV^2 \tag{7.4}$$

$$dV = i * \frac{dt}{C_{tot}} + i * R_{tot} \tag{7.5}$$

The amount of energy a capacitor is capable of storing can be increased by either increasing the capacitance or the voltage stored on the capacitor. The stored voltage is limited by the voltage-withstand-strength of the dielectric (which impacts the distance between the plates). Capacitance can be increased by increasing the area of the plates, increasing the permittivity, or decreasing the distance between the plates. As with batteries, the turnaround efficiency when charging and discharging capacitors is also an important consideration, as is response time. The effective series resistance (ESR) of the capacitor has a significant impact on both. The total voltage change when charging or discharging capacitors is shown in (7.5). Note that C_{tot} and R_{tot} are the result from a combined series/parallel configuration of capacitor cells to increase the total capacitance and the voltage level. The product $C_{tot}R_{tot}$ determines the response time of the capacitor for charging or discharging.

Capacitors are used in many AC and DC applications in power systems. DC storage capacitors can be used for energy storage for power applications. They have long seen use in pulsed power applications for high-energy physics and weapons applications. However, the present generation of DC storage capacitors sees limited use as large-scale energy storage devices for power systems. Capacitors are often used for very short-term storage in power converters. Additional capacitance can be added to the DC bus of motor drives and consumer electronics to provide added ability to ride voltage sags and momentary interruptions. The main transmission or distribution system application where conventional DC capacitors are used as large-scale energy storage is in the distribution dynamic voltage restorer (DVR), a custom power device that compensates for temporary voltage sags on distribution systems. The power converter in the DVR injects sufficient voltage to compensate for the voltage sag, such that loads connected to the system are isolated from the sag. The DVR uses energy stored in DC capacitors to supply a component of the real power needed by the load.

Several varieties of advanced capacitors are in development, with several available commercially for low power applications. These capacitors have significant improvements in one or more of the following characteristics: higher permittivities, higher surface areas, or higher voltage-withstand capabilities. Ceramic hypercapacitors have both a fairly high voltage-withstand (about 1 kV) and a high dielectric strength, making them good candidates for future storage applications. At present, they are largely used in low-power applications. In addition, hypercapacitors have low effective-series-resistance values. Cryogenic operation appears to offer significant performance improvements. The combination of higher voltage-withstand and low effective-series-resistance will make it easier to use hypercapacitors in high-power applications with simpler configurations possible.

Ultracapacitors (also known as supercapacitors) are double-layer capacitors that increase energy storage capability due to a large increase in surface area through use of a porous electrolyte (they still have relatively low permittivity and voltage-withstand capabilities). Several different combinations of electrode and electrolyte materials have been used in ultracapacitors, with different combinations resulting in varying capacitance, energy density, cycle life, and cost characteristics. At present, ultracapacitors are most applicable for high peak-power, low-energy situations. Capable of floating at full charge for 10 years, an ultracapacitor can provide extended power availability during voltage sags and momentary interruptions. Ultracapacitors can be stored completely discharged and installed easily, are compact in size, and can operate effectively in diverse (hot, cold, and moist) environments. Ultracapacitors are now available commercially at lower power levels.

As with BESSs, application of capacitors for power applications will be influenced by the ability to charge and discharge the storage device.

At present, ultracapacitors and hypercapacitors have seen initial application in low-energy applications with much of the development for higher-energy applications geared toward electric vehicles. Near-term applications will most likely use these capacitors in power quality applications. For example, ultracapacitors can be added to the DC bus of motor drives to improve ride-through times during voltage sags. Ultracapacitors can also be added to a DVR or be interfaced to the DC bus of a distribution static compensator (DStatCom).

7.2.4 Flywheel Energy Storage (FES)

Flywheels can be used to store energy for power systems when the flywheel is coupled to an electric machine. In most cases, a power converter is used to drive the electric machine to provide a wider operating range. Stored energy depends on the moment of inertia of the rotor and the square of the rotational velocity of the flywheel, as shown in (7.6). The moment of inertia (I) depends on the radius, mass, and height (length) of the rotor, as shown in (7.7). Energy is transferred to the flywheel when the machine operates as a motor (the flywheel accelerates), charging the energy storage device. The flywheel is discharged when the electric machine regenerates through the drive (slowing the flywheel).

$$E = \frac{1}{2}I\omega^2 \tag{7.6}$$

$$I = \frac{r^2 mh}{2} \tag{7.7}$$

The energy storage capability of flywheels can be improved by increasing the moment of inertia of the flywheel or by turning it at higher rotational velocities or both. Some designs use hollow cylinders for the rotor allowing the mass to be concentrated at the outer radius of the flywheel, improving storage capability with a smaller weight increase.

Two strategies have been used in the development of flywheels for power applications. One option is to increase the inertia by using a steel mass with a large radius, with rotational velocities up to approximately 10,000 rpm. A fairly standard motor and power electronic drive can be used as the power conversion interface for this type of flywheel. Several restorer flywheels using this type of design are available commercially as uninterruptible power supplies (UPSs). This design results in relatively large and heavy flywheel systems. Rotational energy losses will also limit the long-term storage ability of this type of flywheel.

The second design strategy is to produce flywheels with a light-weight rotor turning at very high rotational velocities (up to 100,000 rpm). This approach results in compact, lightweight energy storage devices. Modular designs are possible, with a large number of small flywheels possible as an alternative to a few large flywheels. However, rotational losses due to drag from air and bearing losses result in significant self-discharge, which poses problems for long-term energy storage. High-velocity flywheels are therefore operated in vacuum vessels to eliminate air resistance. The use of magnetic bearings helps improve the problems with bearing losses. Several projects are developing superconducting magnetic bearings for high-velocity flywheels. The near elimination of rotational losses will provide flywheels with high charge and discharge efficiency. The peak power transfer ratings depend on the power ratings in the power electronic converter and the electric machine. Flywheel applications under consideration include automobiles, buses, high-speed rail locomotives, and energy storage for electromagnetic catapults on next-generation aircraft carriers. The high rotational velocity also results in the need for some form of containment vessel around the flywheel in case the rotor fails mechanically. The added weight of the containment can be especially important in mobile applications. However, some form of containment is necessary for stationary systems as well. The largest commercially available flywheel system is about 5 MJ/1.6 MVA weighing approximately 10,000 kg. Flywheel energy storage can be implemented in several power system applications. If an FES system is included with a FACTS or custom power device with a DC bus, an inverter is added to couple the flywheel motor or generator to the DC bus. For example, a flywheel based on an AC machine could have an inverter interface to the DC bus of the custom power device, as shown in Figure 7.4. Flywheel energy storage has been considered for several power system applications, including power quality applications as well as peak shaving and stability enhancement.

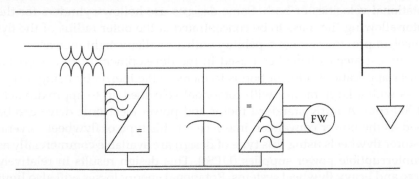

Figure 7.4 Flywheel energy storage coupled to a dynamic voltage.

7.2.5   Pumped Hydroelectric Energy Storage

Hydroelectric storage is a process that converts electrical energy to potential energy by pumping water to a higher elevation, where it can be stored indefinitely and then released to pass through hydraulic turbines and generate electrical energy. A typical pumped-storage development is composed of two reservoirs of equal volume situated to maximize the difference in their levels. These reservoirs are connected by a system of waterways along which a pumping generating station is located. Under favorable geological conditions, the station will be located underground; otherwise, it will be situated on the lower reservoir. The principal equipment of the station is the pumping-generating unit, which is generally reversible and used for both pumping and generating, functioning as a motor and pump in one direction of rotation and as a turbine and generator in opposite rotation.

7.2.6   Flow Batteries

Flow batteries (FBs) are a promising technology that decouples the total stored energy from the rated power. The rated power depends on the reactor size, whereas the stored capacity depends on the auxiliary tank volume. These characteristics make the FB suitable for providing large amounts of power and energy required by electrical utilities. FBs work in a similar way as hydrogen fuel cells (FCs), as they consume two electrolytes that are stored in different tanks (no self-discharge), and there is a microporous membrane that separates both electrolytes but allows selected ions to cross through, creating an electrical current. There are many potential electrochemical reactions, usually called reduction–oxidation reaction (REDOX), but only a few of them seem to be useful in practice [45].

Figure 7.5 shows a schematic of an FB. The power rating is defined by the flow reactants and the area of the membranes, whereas the electrolyte tank capacity defines the total stored energy. In a classical battery the electrolyte is stored in the cell itself, so there is a strong coupling between the power and energy rating. In the cell (flow reactor), a reversible electrochemical reaction takes place, producing (or consuming) electric DC current. At this time, several large- and small-scale demonstration and commercial products use FB technology.

The main advantages of FB technology are (1) high power and energy capacity; (2) fast recharge by replacing exhaust electrolyte; (3) long life enabled by easy electrolyte replacement; (4) full discharge capability; (5) use of nontoxic materials; and (6) low-temperature operation. The main disadvantage of the system is the need for moving mechanical parts such as pumping systems that make system miniaturization difficult. Therefore, the commercial uptake to date has been limited.

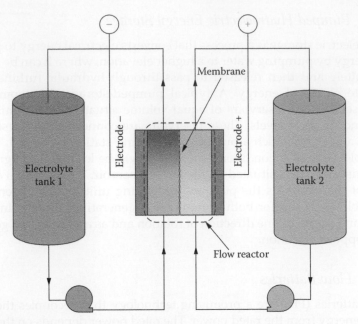

Figure 7.5 (See color insert.) Flow battery cell.

7.2.7 Compressed Air Energy Storage

Compressed air energy storage (CAES) is a technology that stores energy as compressed air for later use. Energy is extracted using a standard gas turbine, where the air compression stage of the turbine is replaced by the CAES, thus eliminating the use of natural gas fuel for air compression. System design is complicated by the fact that air compression and expansion are exothermic and endothermic processes, respectively. With this in mind, three types of systems are considered to manage the heat exchange:

1. Isothermal storage, which compresses the air slowly, thus allowing the temperature to equalize with the surroundings. Such a system works well for small systems where power density is not paramount.
2. Adiabatic systems, which store the released heat during compression and feed it back into the system during air release. Such a system needs a heat-storing device, complicating the system design.
3. Diabatic storage systems, which use external power sources to heat or cool the air to maintain a constant system temperature. Most commercially implemented systems are of this kind due to high power density and great system flexibility, albeit at the expense of cost and efficiency.

CAES systems have been considered for numerous applications, most notably for electric grid support for load leveling applications. In such systems, energy is stored during periods of low demand and then converted back to electricity when the electricity demand is high. Commercial systems use natural caverns as air reservoirs to store large amounts of energy; installed commercial system capacity ranges from 35 to 300 MW.

7.2.8   Thermoelectric Energy Storage

Thermoelectric energy storage (TEES) for solar thermal power plants consists of a synthetic oil or molten salt that stores energy in the form of heat collected by solar thermal power plants to enable smooth power output during daytime cloudy periods and to extend power production for 1–10 h after sunset. End-use TEES stores electricity from off-peak periods through the use of hot or cold storage in underground aquifers, water or ice tanks, or other storage materials and uses this stored energy to reduce the electricity consumption of building heating or air conditioning systems during times of peak demand.

7.2.9   Hybrid Energy Storage Systems

Certain applications require a combination of energy, power density, cost, and life cycle specifications that cannot be met by a single energy storage device. To implement such applications, hybrid energy storage devices (HESDs) have been proposed. HESDs electronically combine the power output of two or more devices with complementary characteristics. HESDs all share a common trait: combining high-power devices (devices with quick response) and high-energy devices (devices with slow response). Proposed HESDs are listed next, with the energy-supplying device listed first followed by the power-supplying device:

1. Battery and electric double-layer capacitor (EDLC)
2. FC and battery or EDLC
3. CAES and battery or EDLC
4. Battery and flywheel
5. Battery and SMES

7.3   Energy Storage Systems Compared

There are relative advantages and disadvantages of the various energy storage systems. For example, some of the disadvantages of BESSs include limited life cycle, voltage and current limitations, and potential environmental hazards. Again, some of the disadvantages of pumped hydroelectric are large unit sizes and topographic and environmental limitations. The major problems of confronting the implementation of SMES units are the high cost and environmental issues associated with strong magnetic field. Relatively short duration, high frictional loss (windage), and low-energy density restrain the flywheel systems from the application in energy management. Similar to flywheels, the major problems associated with capacitors are the short durations and high-energy dissipations due to self-discharge loss. However, among all energy storage systems, SMES is the most effective from the viewpoints of fast response, charge and discharge cycles, and control ability of both active and reactive powers simultaneously.

7.4   Using SMES to Minimize Fluctuations in Power, Frequency, and Voltage of Wind Generator Systems

Recent developments and advances in both superconducting and power electronics technology have provided the power transmission and distribution industry with SMES systems. SMES is a large superconducting coil capable of storing electric energy in the magnetic field generated by DC current flowing through it. The real power as well as the reactive power can be absorbed (charging) by or released (discharging) from the SMES coil according to system power requirements. Since the successful commissioning test of the Bonneville Power Administration (BPA) 30 MJ unit, SMES systems have received much attention in power system applications, such as diurnal load demand leveling, frequency control, automatic generation control, and uninterruptible power supplies. A gate turn-off (GTO) thyristor-based SMES unit is able to absorb and inject active as well as reactive powers simultaneously in rapid response to power system requirements. Therefore, it can act as a good tool to decrease voltage and frequency fluctuations of the system considerably. With this view, minimization of fluctuations of line power and terminal voltage of wind generators by the SMES is considered in this book.

The model system as shown in Figure 7.6 was used for the simulation [2]. The power system model belongs to Ulleung Island in South Korea. The model system consists of two diesel generators (4.5 MVA and 1.5 MVA), two hydroelectric generators (0.6 MVA and 0.1 MVA), one wind turbine generator (0.6 MVA IG), and a load of 6 MW. In this work, to evaluate the performance of SMES systems in detail, another load of 2 MW is also considered. When a 2 MW load is considered, ratings of some of the generators and transformers are changed, as shown in red colors in Figure 7.6. A condenser C is connected to the terminal of the wind generator to compensate the reactive power demand for the induction generator at steady-state. The value of C has been chosen so that the power factor

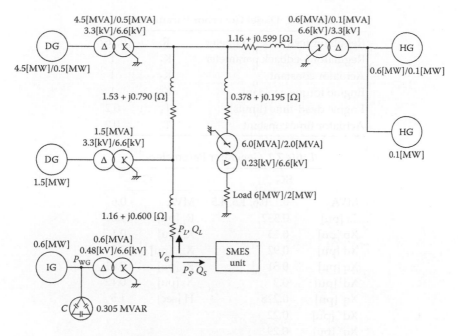

Figure 7.6 (See color insert.) Power system model.

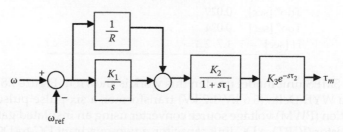

Figure 7.7 Diesel–governor model.

of the wind power station becomes unity when it is generating the rated power (P = 0.6, V = 1.0). The SMES unit is located at the induction generator terminal bus. The automatic voltage regulator (AVR) control system models for the diesel and hydraulic generators and the governor (GOV) control system model for the hydroelectric generator used in this work are the built-in models of Power System Computer Aided Design (PSCAD)/ Electromagnetic Transients in DC (EMTDC). However, the GOV control system model for the diesel generator used in this work is shown in Figure 7.7. The typical values for the parameters of the diesel governor model are shown in Table 7.1, whereas synchronous generator parameters as well as induction generator parameters used for this simulation are shown in Table 7.2.

Table 7.1 Diesel Governor Parameters

Regulator feedback parameter	R	0.167
Regulator feedback parameter	K_1	1
Actuator constant	K_2	1
Engine torque constant	K_3	1
Engine dead-time (limits)	T_1	0.2
Actuator time constant	T_2	0.2

Table 7.2 Generator Parameters

SG		IG	
MVA	0.1, 0.6, 1.5, 4.5	MVA	0.6
Ta [pu]	0.332	R_1 [pu]	0.01
Xp [pu]	0.13	X_1 [pu]	0.1
Xd [pu]	0.92	X_{mu} [pu]	3.5
Xq [pu]	0.51	R_2 [pu]	0.01
Xd' [pu]	0.3	X_2 [pu]	0.12
Xq' [pu]	0.228	H [sec]	1.5
Xd" [pu]	0.22		
Xq" [pu]	0.29		
Tdo' [sec]	5.2		
Tdo" [sec]	0.029		
Tqo" [sec]	0.034		
H [sec]	1.7, 2.3		

The SMES unit model used in this work is shown in Figure 7.8. It consists of a WYE-Delta (6.6 kV/1.2 kV) transformer, a six-pulse pulse width modulation (PWM) voltage source converter using an insulated gate bipolar transistor (IGBT), a DC link capacitor, a two-quadrant DC-to-DC chopper using IGBT, and an inductance as a superconducting coil. The VSC and the DC-to-DC chopper are linked by a DC link capacitor. In this work, to evaluate in detail the performance of the SMES system to minimize frequency fluctuations, different energy capacities of the SMES are considered. The parameters for the proposed SMES are shown in Table 7.3.

The PWM VSC provides a power electronic interface between the AC power system and the superconducting coil. The DC link voltage E_{DC} and grid point voltage V_G are maintained as constant by the VSC. In this model system, the transformer connecting the VSC and the AC system is expressed by an RL circuit. Since a leakage reactance of a transformer is, in general, much greater than the winding resistance, the active and reactive powers of the SMES system are proportional to the d- and q-axis currents and thus also to the d- and q-axis voltages as expressed by (7.8). Based on this concept, the control system of the VSC is constructed.

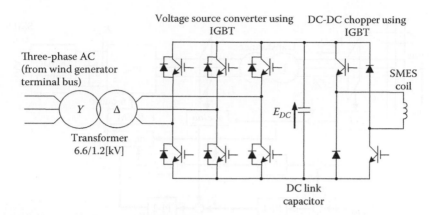

Figure 7.8 Configuration of SMES unit.

Table 7.3 SMES Parameters

SMES capacity (MW)		0.33	
Energy capacity (MJ)	1	5	10
Rated current (kA)	0.5	0.82	1
Inductance (H)	8	15	20
DC link capacitor (mF)		10	

$$\begin{cases} P_S \propto I_d \propto -V_q \\ Q_S \propto -I_q \propto -V_d \end{cases} \tag{7.8}$$

The control system of the VSC is shown in Figure 7.9. The proportional-integral (PI) controllers determine reference d- and q-axis currents by using the difference between the DC link voltage E_{DC} and reference value $E_{DC\text{-}ref}$ and the difference between terminal voltage V_G and reference value $V_{G\text{-}ref}$ respectively. The reference signal for the VSC is determined by converting d- and q-axis voltages obtained from the difference between reference d–q-axes currents and their detected values. Parameters of the Proportional-Integral (PI) controllers are determined by trial and error method. The PWM signal is generated for IGBT switching by comparing the reference signal, which is converted to a three-phase sinusoidal wave with the triangular carrier signal.

The superconducting coil is charged or discharged by adjusting the average voltage, $V_{sm\text{-}av}$, across the coil, which is determined by the duty cycle of the two-quadrant DC-to-DC chopper. Based on this concept, the control system of a two-quadrant DC-to-DC chopper is constructed as shown in Figure 7.10. The duty cycle is determined by the PI controller. For

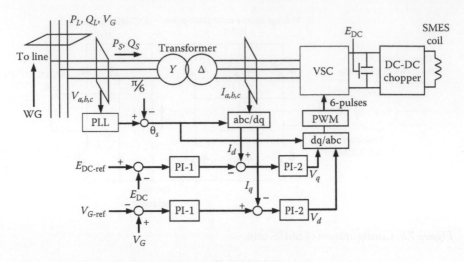

Figure 7.9 Control system of the VSC.

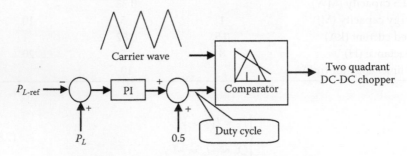

Figure 7.10 Control system of two-quadrant DC-DC chopper.

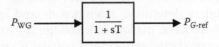

Figure 7.11 Determination of reference value.

example, when the duty cycle is larger than 0.5 or less than 0.5, the stored energy of the coil is either charging or discharging. To generate the PWM gate signals for the IGBT of the chopper, the reference signal is compared with the triangular signal.

The reference value of the transmission line power P_{G-ref} is determined by a low-pass filter (LPF) as shown in Figure 7.11. LPF consists of a first-order delay system. Though it is very simple, a reference value with enough smoothing effect can be obtained by this type of LPF.

7.4.1 Method of Calculating Power System Frequency

In this study, the index of the smoothing effect is used in power system frequency analysis. Power system frequency fluctuation occurs due to an imbalance between supply and load power in power system. Then, the frequency fluctuation can be described using two components: the rate of generator output variation, K_G (MW/Hz), and load variation, K_L (MW/Hz). They are representing the amount of power variation causing 1 Hz frequency fluctuation. When generator output variation, ΔG (MW), and load variation, ΔL (MW), occur, frequency fluctuation of the power system, ΔF (Hz), is expressed as

$$\Delta F = \frac{\Delta G - \Delta L}{K_G + K_L} \qquad (7.9)$$

$$K = K_G + K_L \qquad (7.10)$$

where K is frequency characteristic constant. In general, frequency characteristic is expressed as percentage K_G (expressed as %K_G) for the total capacity of all generators and percentage K_L (expressed as %K_L) for the total load. In general, it is known that %K_G and %K_L are almost constant and generally take a value of 8–15% MW/Hz and 2–6% MW/Hz, respectively. However, K_L and K_G change greatly during a day because the number of parallel generators changes depending on the amount of load during a day. And when power imbalance ΔP occurs in power system, frequency fluctuation $\Delta P/K$ cannot occur immediately due to the governor characteristic and generator inertia. Normally, ΔF converges to a new steady-state value in 2 to 3 sec. In general, when ΔP is changing slowly, the relationship between ΔP and ΔF can be expressed as follows:

$$\frac{\Delta F}{\Delta P} = \frac{1}{K(1+sT)} \qquad (7.11)$$

where $\Delta P = \Delta G - \Delta L$. Since changing load is not considered in this study, ΔL is 0. Time constant, T (sec), depending on the setting of generator governor and generator inertia, is generally 3 to 5 sec. In this study, power system capacity is assumed to be 100 MW, and frequency characteristic K (MW/Hz) is selected to 8 MW/Hz. This selection means that adjustability of the system frequency is weak, resulting in a severe situation. Similarly, time constant T is selected to 3 sec. In this study, frequency fluctuation in the power system is evaluated using Equation (7.11). Therefore, frequency fluctuation, ΔF, is obtained as shown in Figure 7.12.

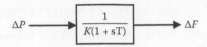

Figure 7.12 Frequency calculation model.

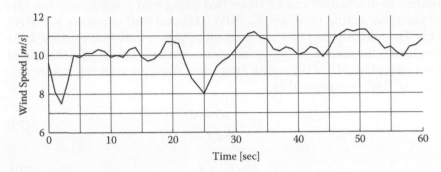

Figure 7.13 Wind speed data.

Table 7.4 Specifications of Wind Speed

Average [m/s]	10.1
Minimum [m/s]	7.3
Maximum [m/s]	11.3
Standard deviation [m/s]	0.79

7.4.2   Simulation Results and Discussions

To evaluate the performance of the SMES strategy to minimize system frequency fluctuations, simulations are carried out considering variable wind speed data as shown in Figure 7.13. The specifications of wind speed are given in Table 7.4. The simulation time and time step were chosen as 60 sec and 10 msec, respectively. Figures 7.14 and 7.15 show the responses of active powers of the wind generators, diesel generators and hydraulic generators, and the system frequency without SMES systems with 2 MW and 6 MW loads, respectively. It is seen that without the wind generator the system frequency is maintained at the nominal value of 60 Hz; however, when the wind generator is used in the power system model the system frequency considerably fluctuates, especially for low load (2 MW load). This fact motivates the use of the SMES method to minimize system frequency and power fluctuations resulting from the wind generator system.

7.4.2.1   Effectiveness of SMES Systems on Minimizing Wind Generator Power, Frequency, and Voltage Fluctuations

Figures 7.16 and 7.17 show the responses of active powers of the wind generator, diesel generators and hydraulic generators, the transmission line

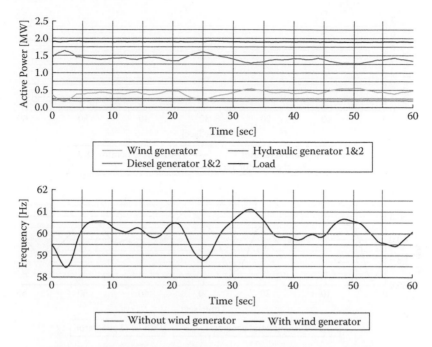

Figure 7.14 (See color insert.) Responses of active power and system frequency without SMES (Load 2 MW).

power, and the system frequency using 10 MJ SMES with 2 MW and 6 MW loads, respectively. It is seen that because of the use of the SMES, the system frequency fluctuations are minimized well, and the frequency is maintained almost at the level of the nominal value of 60 Hz for both loads. Also, it is observed that the SMES system can successfully smooth the transmission line power for both loads. Thus, the smoothed power can be supplied to the consumers. Although the grid voltage responses are not shown here, it is seen that the SMES system can minimize the oscillations of grid voltage. As a whole, the SMES system can be considered a very effective tool to minimize frequency, power, and voltage fluctuations of power systems including wind generators.

7.4.2.2 Comparison among Energy Capacities of SMES Systems to Minimize Wind Generator Power, Frequency, and Voltage Fluctuations

In this work, to evaluate in detail the performance of SMES systems to minimize frequency, power, and voltage fluctuations, extensive simulations were carried out considering different energy capacities of SMES. Figure 7.18 shows the responses of active power of a wind generator, transmission line power, and system frequency using 10 MJ, 5MJ, and 1 MJ

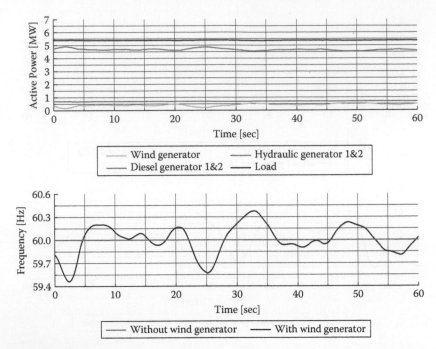

Figure 7.15 (See color insert.) Responses of active power and system frequency without SMES (Load 6 MW).

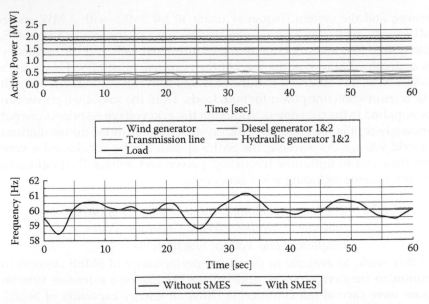

Figure 7.16 (See color insert.) Responses of active power and system frequency with 10 MJ SMES (Load 2 MW).

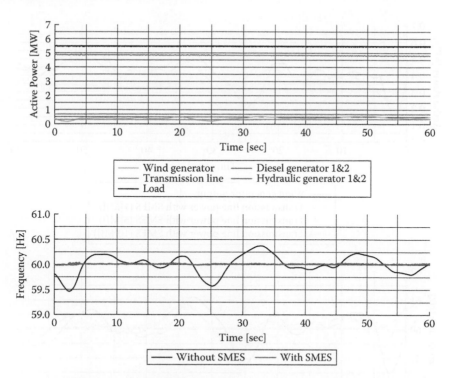

Figure 7.17 (See color insert.) Responses of active power and system frequency with 10 MJ SMES (Load 6 MW).

SMES with a 2 MW load. It is seen that the performance of the 10 MJ SMES is the best from the viewpoint of minimization of fluctuations of system frequency and line power. The 5 MJ SMES can minimize the power and frequency fluctuations well; however, the 1 MJ SMES has little ability to minimize frequency and power fluctuations. In general, the larger the capacity of SMES is, the better the ability of fluctuations minimization becomes. However, if the large size of SMES is adopted, its cost increases, making the installation of SMES impractical. From this viewpoint, the 5 MJ SMES is a trade-off between higher cost and better performance. Therefore, the SMES capacity for fluctuations minimization of power, frequency, and voltage should be selected considering the viewpoint of cost-effectiveness.

Figures 7.19, 7.20, and 7.21 show the responses of real power, storage energy, coil current, coil voltage, DC link current, and DC link voltage with 2 MW and 6 MW loads corresponding to 5 MJ, 10 MJ, and 1 MJ SMES systems, respectively. In all cases it is seen that the SMES is being charged and discharged according to system power requirements to minimize frequency fluctuations. The SMES energy and coil current are well within the ranges of their rated values. Moreover, the PWM voltage source converter can maintain the DC link voltage constant.

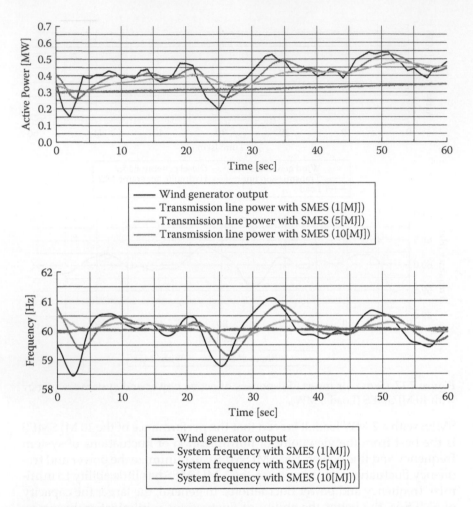

Figure 7.18 (See color insert.) Responses of active power and system frequency with different capacities of SMES (Load 2 MW).

7.4.3   SMES Power and Energy Ratings

It is important to know the optimal power and energy ratings of SMES systems so that their cost is minimal. In this work, the relationship between SMES power capacity and the smoothing ability is investigated by evaluating a wind turbine generator output, P_{WG}, and reference value of transmission line power, $P_{G\text{-ref}}$, with the condition that the LPF time constant is changed from 3 to 300 sec. Since a large number of wind turbine generators are going to be connected to a power system in the near future, a percentage of wind turbine generators to the total power system capacity is assumed fairly large: 10% (10 MW) and 20% (20 MW) in this

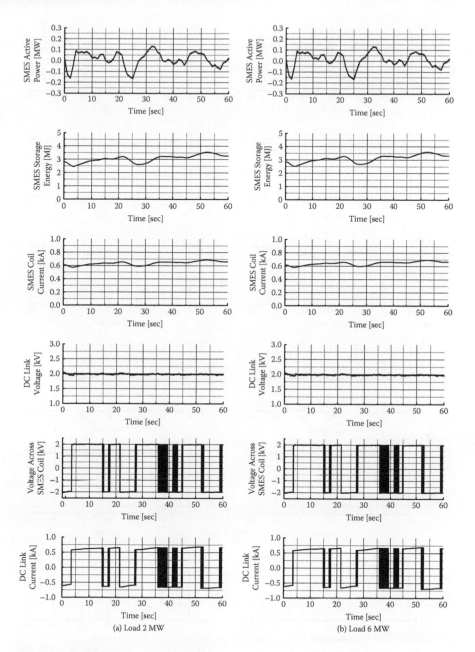

Figure 7.19 Responses of 5 MJ SMES.

study. Smoothing of the wind turbine generator output is investigated by using $P_{G\text{-ref}}$ instead of an SMES unit in the calculation to estimate a required capacity of the SMES [3].

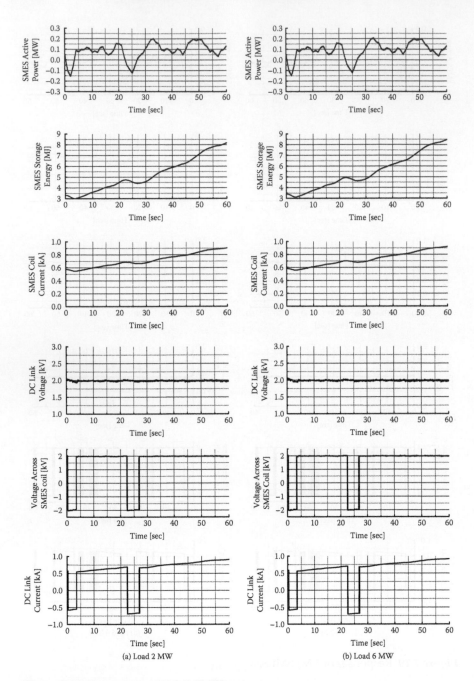

Figure 7.20 Responses of 10 MJ SMES.

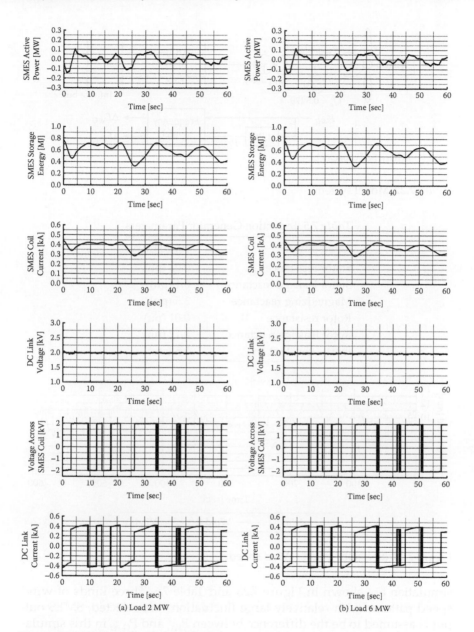

Figure 7.21 Responses of 11 MJ SMES.

The model system used in this simulation analysis is shown in Figure 7.22. Table 7.5 shows the parameters for the induction generator shown in Figure 7.22. The wind speed patterns and conditions used in the

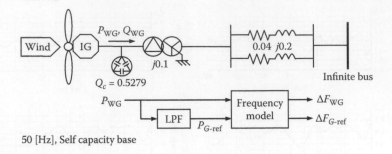

Figure 7.22 Model system.

Table 7.5 Induction Generator Parameters

Rated power	10, 20 (MW)
Stator resistance	0.01 (pu)
Stator leakage reactance	0.1 (pu)
Magnetizing reactance	3.5 (pu)
Rotor resistance	0.01 (pu)
Rotor leakage reactance	0.12 (pu)
Inertia constant (H)	1.5 (sec)

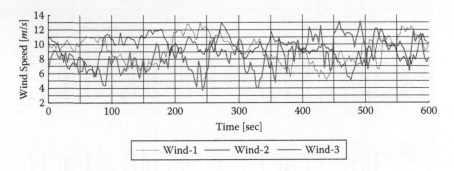

Figure 7.23 (See color insert.) Wind speed.

simulation are shown in Figure 7.23 and Table 7.6. Three kinds of wind speed patterns with relatively large fluctuations are selected. SMES output is assumed to be the difference between P_{WG} and P_{G-ref} in this simulation, and then a standard deviation of the SMES output, σ, is calculated. In addition, smoothing effect is evaluated by using frequency fluctuation ΔF (ΔF_{WG}, ΔF_{G-ref}). ΔF is calculated by applying P_{WG} and P_{G-ref} to the frequency calculation model. The required SMES power capacity for smoothing P_{WG} and an LPF time constant suitable for the reference value with enough smoothing effect are investigated by using σ and ΔF in this simulation.

Table 7.6 Wind Speed Condition

Wind Data's Name	Average Wind Speed [m/s]		Standard Deviation of Wind Speed [m/s]	
Wind-1	Medium	9.28	Large	1.82
Wind-2	Medium	8.45	Large	2.01
Wind-3	Medium	9.44	Medium	1.54

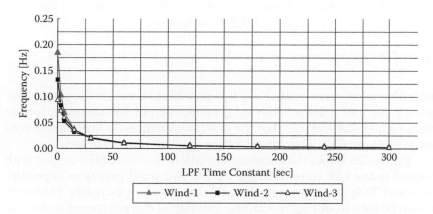

Figure 7.24 (See color insert.) Maximum frequency fluctuation (WG capacity 10%).

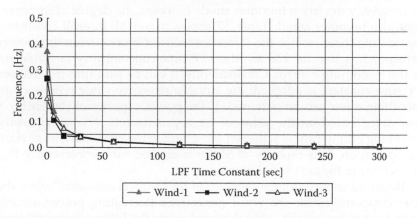

Figure 7.25 (See color insert.) Maximum frequency fluctuation (WG capacity 20%).

Figures 7.24 and 7.25 show the maximum frequency fluctuation with respect to the LPF time constant for two cases of wind generator capacity, 10% and 20%, respectively. Frequency fluctuation decreases as the LPF time constant increases. Therefore, if the transmission line power is compensated

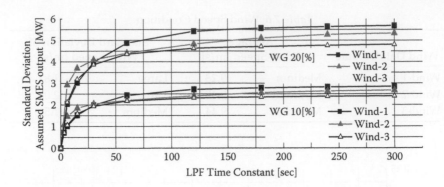

Figure 7.26 (See color insert.) Assumed SMES output standard deviation.

according to reference value $P_{G\text{-ref}}$, it is possible to decrease the system frequency fluctuation due to the wind generator output fluctuations. It is clear from Figures 7.24 and 7.25 that the maximum frequency fluctuations converge to almost 0 Hz when the LPF time constant is over 120 sec.

Figure 7.26 shows the standard deviation, σ, of the SMES output with respect to the LPF time constant for the two wind generator capacities (10% and 20%). σ increases as the LPF time constant increases. However, as can be seen from Figure 7.26, the function of σ is not monotonous and saturates where the LPF time constant is about 120 sec. Therefore, if a longer LPF time constant and thus the larger capacity of SMES are adopted, the frequency deviation becomes small; however, the degree of improvement also becomes small. Table 7.7 shows σ and ΔF in each condition. Considering these results, the reference value of transmission line power corresponding to σ for 120 sec, the LPF time constant may be sufficient for the smoothing control. Consequently, it can be said that if a 120 sec LPF time constant is adopted, the suitable reference value with enough smoothing effect can be obtained. If the power capacity of the SMES is determined based on the value of 2σ, approximately 95% of necessary smoothing output can be achieved according to the characteristic of standard deviation. This capacity of SMES is applied for compensating P_{WG} fluctuations in the next simulation analysis.

Required energy storage capacity of the SMES is estimated when the latter compensates for the wind generator's fluctuating power according to reference power PG-ref calculated using the LPF with several time constants including the best one determined in the previous section. Figure 7.27 shows the model for this evaluation. The wind power output is assumed as an ideal periodic sinusoidal wave form or trapezoidal wave form, and then the behavior of the SMES stored energy is investigated. To evaluate the influence of the period of wind power fluctuation on the

Table 7.7 The Standard Deviation of SMES Output and Maximum Frequency Fluctuation in Each Condition

LPF time const [sec]	Wind Generator Capacity 10 [MW]						Wind Generator Capacity 20 [MW]					
	Wind-1		Wind-2		Wind-3		Wind-1		Wind-2		Wind-3	
	σ [MW]	Δf [Hz]	σ [MW]	Δf [Hz]	σ [MW]	Δf [Hz]	σ [MW]	Δf [Hz]	σ [MW]	Δf [Hz]	σ [MW]	Δf [Hz]
0	0.000	0.133	0.000	0.185	0.000	0.093	0.000	0.266	0.000	0.369	0.000	0.186
15	1.516	0.032	1.848	0.036	1.591	0.036	3.032	0.044	3.697	0.073	3.182	0.072
30	1.985	0.021	2.042	0.019	1.934	0.020	3.971	0.041	4.083	0.038	3.868	0.040
60	2.438	0.011	2.200	0.010	2.176	0.010	4.876	0.022	4.399	0.020	4.352	0.020
120	2.710	0.006	2.411	0.006	2.315	0.005	5.421	0.011	4.823	0.011	4.630	0.010
180	2.788	0.004	2.556	0.004	2.358	0.004	5.575	0.007	5.112	0.008	4.716	0.007
240	2.823	0.003	2.633	0.003	2.383	0.003	5.646	0.006	5.267	0.006	4.765	0.006
300	2.844	0.002	2.669	0.002	2.403	0.002	5.689	0.005	5.338	0.005	4.807	0.004

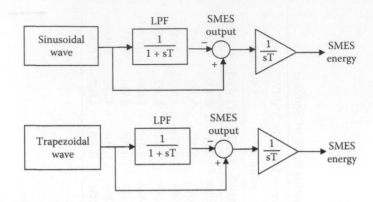

Figure 7.27 Simulation model.

smoothing effect, five LPF time constants (30, 60, 120, 180, and 300 sec) are used and the period of wind power fluctuation is varied from 10 to 1,200 sec. The required energy capacity of SMES can be determined as a difference of the maximum and the minimum values of the stored energy.

Table 7.8 shows a difference of the maximum and the minimum values of the SMES stored energy with respect to the period of fluctuation for several LPF time constants in two cases. In Table 7.8, the result in the case of an extremely long period (12,000 sec) is also indicated. From these results, (7.12) can be established. Consequently, the required SMES energy storage capacity to compensate fluctuating wind power with relatively long periodic change can be decided approximately by using (7.12), and the energy storage capacity greater than this is not needed [3–4].

$$E\ (MJ) = T\ (sec) \times P_{WF}\ (MW) \tag{7.12}$$

where E (MJ) is the saturated value of the range of energy change of the SMES, T (sec) is the LPF time constant, and P_{WF} (MW) is the rated wind farm output.

Since it is shown that the frequency fluctuation can be decreased sufficiently by the reference value $P_{G\text{-ref}}$, which is determined based on a LPF time constant of 120 sec, this value is used to calculate the reference value using the system shown in Figure 7.11. The model system used in the simulation analysis is shown in Figure 7.28. In this simulation analysis, the power capacity of the SMES is determined based on the value of 2σ, because the necessary smoothing effect can be achieved by an SMES with this capacity as shown in the previous section. When the power capacity of the wind generator is 10% of the entire power system, then the SMES power capacity is 5.5 MW ($2\sigma = 2.7 \times 2 = 5.4$, where $\sigma = 2.7$ MW). Similarly, when the wind generator capacity is 20%, the SMES capacity is 11 MW

Table 7.8 The Range of Changing SMES Energy for Each Input

Period (sec)	The Range of Changing SMES Energy (MJ)									
	Sinusoidal-Wave Input					Trapezoidal-Wave Input				
	LPF Time Constant					LPF Time Constant				
	30 (sec)	60 (sec)	120 (sec)	180 (sec)	300 (sec)	30 (sec)	60 (sec)	120 (sec)	180 (sec)	300 (sec)
0	0.00	0.00	0.00	0.00	0.00	0.00	0.00	0.00	0.00	0.00
30	47.20	47.64	47.90	48.52	50.38	69.94	70.91	71.17	71.25	71.85
60	90.98	94.29	95.42	96.72	100.40	132.64	139.87	141.83	142.28	143.55
120	161.11	182.00	189.00	192.25	199.44	221.85	265.29	279.75	282.76	286.23
300	254.02	373.61	444.09	464.91	486.78	294.80	497.67	638.96	677.58	703.14
600	286.20	508.04	747.46	846.21	925.06	299.95	589.60	995.36	1186.59	1330.76
900	293.62	553.41	919.87	1121.20	1298.86	299.99	599.01	1133.44	1493.44	1844.31
1200	296.36	572.41	1016.08	1309.95	1613.18	300.00	599.90	1179.20	1653.86	2220.98
—		-					-			
12000	/	/	1197.63	1792.05	2963.66	/	/	1199.99	1799.99	2999.99

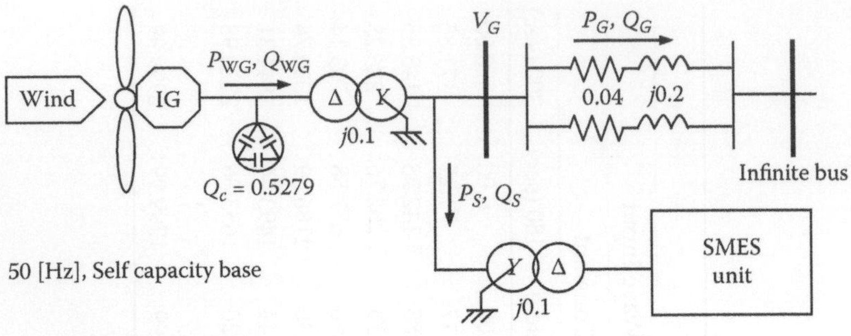

Figure 7.28 Model system.

(σ = 5.4 MW). SMES power capacities in these two cases are both 55% of the wind generator rated capacities (10 and 20% of the entire power system). In the following simulations, two cases using SMES systems with capacities of 55 and 50% of the wind generator rated capacity are examined for the three wind speed patterns.

Figure 7.29 shows responses of wind generator output and transmission line power when the capacity of the wind generator is 10% of the entire power system and the SMES power capacity is 55% and 50% of the wind generator. Figure 7.30 shows responses of frequency fluctuation under the same condition as Figure 7.29. Figure 7.31 shows responses of wind generator output and transmission line power when the capacity of the wind generator is 20% and the SMES power capacity is 55% and 50%. Figure 7.32 shows responses of frequency fluctuation under the same condition as Figure 7.30. Table 7.9 shows values of the maximum frequency fluctuation in each case. As shown in Figures 7.29 to 7.32, since necessary compensating power exceeds the capacity of the SMES system in some instances, large frequency fluctuation occurs, especially in the case of wind-2. It can be seen in Table 7.9 that frequency fluctuations in SMES power capacity at 50% are greater than those in SMES power capacity at 55%. The maximum frequency fluctuation in SMES power capacity at 50% is 0.061 Hz for 10% WG and 0.122 Hz for 20% WG, whereas the maximum frequency fluctuation in SMES power capacity at 55% is 0.006 Hz for 10% WG and 0.012 Hz for 20% WG. The former is approximately 10 times larger than the latter. Therefore, it can be said that the SMES with 50% power capacity of that of wind generator is not sufficient.

Consequently, it can be concluded that if an LPF time constant of 120 sec is selected and an SMES system with 55% power capacity of that of wind generator is adopted, a suitable reference value for the transmission line power can be obtained and then sufficient smoothing effect can be achieved. Moreover, the energy storage capacity of the SMES can be determined to 2,400 MJ according to (7.12).

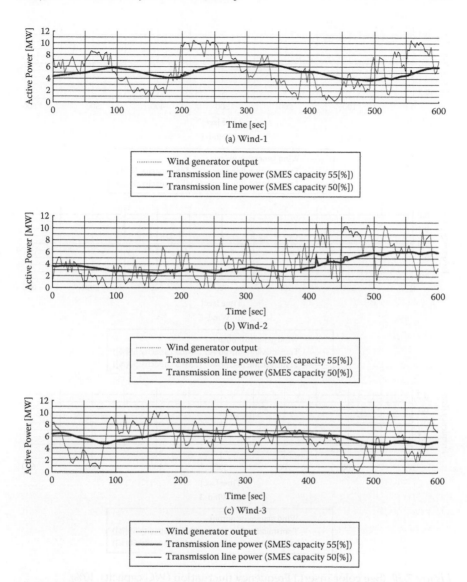

Figure 7.29 (See color insert.) Wind generator output and transmission line power (WG capacity 10%).

7.5 Power Quality Improvement Using a Flywheel Energy Storage System

A flywheel energy storage system using a squirrel-cage induction machine is explained in this context. The system uses the squirrel-cage induction machine, which is widely available and inexpensive, and the simple volt/

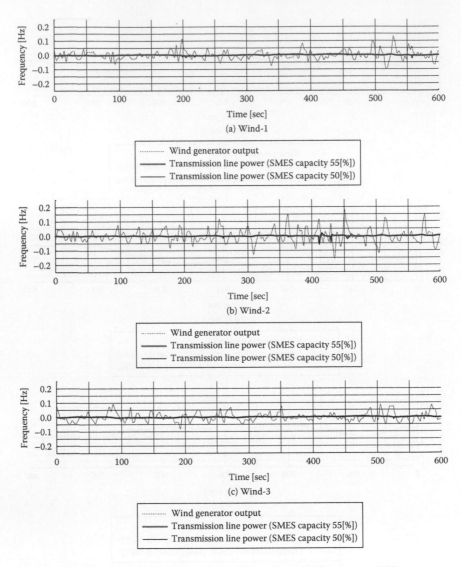

Figure 7.30 (See color insert.) Frequency fluctuation (WG capacity 10%).

hertz control technique with just nameplate data as machine parameters. Therefore, no complex parameter measurement is necessary, and the system has an advantage on parallel operation because adding or replacing units is straightforward. Hence, it can easily operate with different types of storage or distributed energy sources in DC bus microgrid systems. Moreover, the control scheme improves the overall stability of the DC bus system [66].

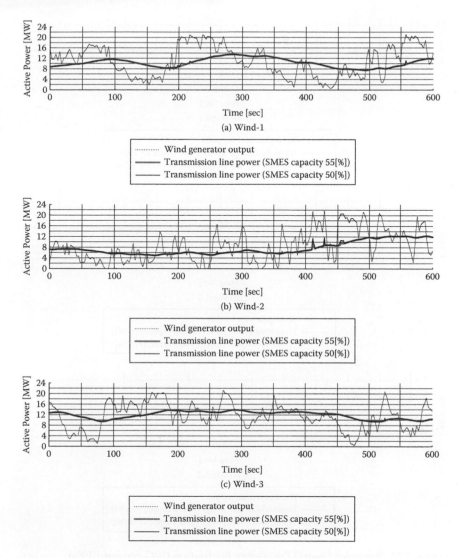

Figure 7.31 (See color insert.) Wind generator output and transmission line power (WG capacity 20%).

Advances in power electronics, magnetic bearings, and flywheel materials have made flywheel systems a viable energy storage option. Although it has higher initial cost than batteries, flywheel energy storage has advantages such as longer lifetime, lower operation and maintenance costs, and higher power density (typically by a factor of 5 to 10). Flywheel systems have been used in many applications instead of or in conjunction with batteries. Machines such as permanent magnet (PM) machines,

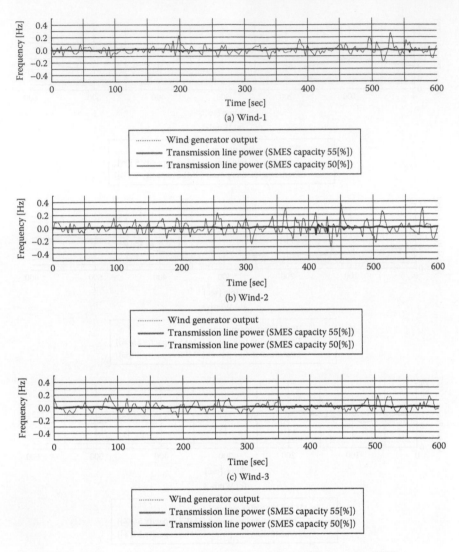

Figure 7.32 (See color insert.) Frequency fluctuation (WG capacity 20%).

synchronous reluctance machines, synchronous homopolar machines, and induction machines have been explored for flywheel motors and generators. PM machines have advantages such as lower rotor losses, high power factor, efficiency, and power density. However, high-power magnets are costly and have an inherent disadvantage of spinning losses.

Synchronous homopolar machines, although they have been researched for various applications, are not widely used in practice. Synchronous reluctance machines can be a viable choice for a flywheel motor or generator, but the machines are not easily available. Induction machine-based

Table 7.9 Maximum Frequency Fluctuation in Each Condition

Wind Generator Capacity	Wind Speed Data	ΔFWG (Hz)	ΔFG [Hz] SMES Power Capacity	
			55 (%)	50 (%)
10 (%)	wind-1	0.133	0.009	0.025
	wind-2	0.184	0.048	0.061
	wind-3	0.092	0.011	0.017
20 (%)	wind-1	0.265	0.019	0.05
	wind-2	0.369	0.095	0.122
	wind-3	0.185	0.024	0.031

flywheel systems have been investigated, and it has been suggested that the rugged and inexpensive induction machines are good candidates for high-power flywheel motors or generators. Field-oriented vector controllers are generally used for faster dynamic response, which require complex machine parameter measurements and complicated controllers.

However, the majority of the fast disturbances are shorter than several seconds, and storage devices needed to store the intermittent generation of renewable energy sources do not necessarily have to be fast if they are not focusing on transient performance improvement. Considering the overall cost, it would be more efficient for a microgrid system to use a combination of faster storage devices for short transient and slower but inexpensive storage units for massive energy charge and discharge with renewable energy sources such as wind turbines and photovoltaic systems.

7.5.1 DC Bus Microgrid System

Among the microgrids that have been researched recently, low-voltage DC (LVDC) bus-based systems have received attention because of advantages such as fast control without need for communication, a variety of DC energy sources and loads, and efficiencies on system size and cost. A block diagram of a typical DC bus microgrid system is shown in Figure 7.33.

The LVDC systems use the voltage droop technique, involving the DC bus voltage as a command signal. DC bus systems do not have the functional issues of AC systems such as synchronization and reactive power compensation. Unlike large-scale distribution systems, the LVDC microgrid system does not have a resistive loss issue because the length of the bus is much shorter. Also, sharing a DC bus has a structural advantage because all the energy sources and loads are connected via DC-to-DC or DC-to-AC voltage source converters that use DC voltage as their medium. Another layer of DC-to-AC converters is necessary for some subsystems to make an AC bus system.

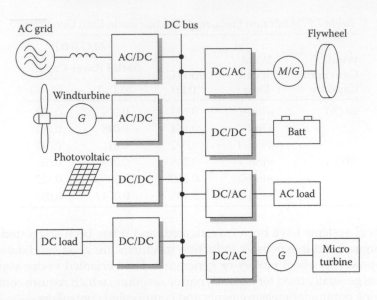

Figure 7.33 Conceptual diagram of a DC bus microgrid system.

Using the fast-acting power electronics converters, constant voltage can be supplied to the loads regardless of some fluctuations on the bus side, and the renewable energy sources can readily generate maximum power with maximum power point tracking (MPPT) techniques. As a small-scale power system, a microgrid can have relatively higher load fluctuations, especially when it is not operating in grid-connected mode. This is because the inertia of the generators is not as large as that found in the large-scale synchronous generators. Generation of power from the renewable energy sources relies heavily on natural conditions, and they are intermittent. Hence, energy storage devices are required for stable operation of a microgrid system in either grid-connected or islanded operations. A flywheel-based energy storage system is investigated in this work.

To develop a cost-effective system, a squirrel-cage induction machine is selected for the motor or generator of the flywheel energy storage in this work. A per-phase equivalent circuit of an induction machine is shown in Figure 7.34. Easy parallel operation is an important factor in energy storage for higher capacity.

7.5.2   Volt/Hertz Control

Volt/hertz control has been widely used for induction machine speed control. The field-oriented vector control technique has been used for applications that need fast response, but it requires parameter measurements, current feedback, a machine model, and controller tuning. On the

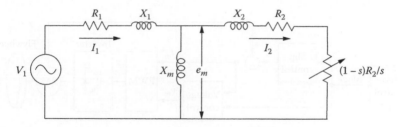

Figure 7.34 Induction machine per-phase equivalent circuit.

contrary, all of the information necessary to run an induction machine in rated condition, such as voltage, speed, and slip, can be found on the nameplate of the machine for volt/hertz control.

Much research on flywheel energy storage systems has used field-oriented controllers because the applications need quite fast energy flow, for example, Uninterruptable Power Supply (UPS) application compensating the voltage dip. However, if an energy storage system absorbs or releases the energy in slow dynamics, volt/hertz control can be a valid candidate due to its simplicity and inherent stability.

7.5.3 Microgrid System Operation

The microgrid system shown in Figure 7.33 uses the DC bus voltage as a control signal. Hence, a prime mover, such as a grid-connected converter or micro turbine unit, does not control the voltage tightly at a fixed reference voltage to reflect the energy flow on the bus voltage if it is in the nominal operating region above the minimum threshold. All of the renewable energy source units are operating in the power mode, which is generating its maximum power when the energy is available. Hence, the bus voltage will rise above the nominal operating range if the generation is larger than the load power consumption. The storage devices detect the excessive energy in the bus and absorb it. If the generated energy is large enough to exceed the maximum threshold, the grid-connected converter can push it back to the grid.

When the DC bus voltage gets lower than the discharge threshold due to the increased load or decreased generation, the energy stored in the storage units is discharged to the bus to maintain the bus voltage at the minimum level. If all the energy storages and generators are not able to hold the bus voltage at the minimum level, load shedding can be initiated by disconnecting some of the power electronics converters supporting lower priority loads. If different kinds of storage or load control units are connected to the bus, the priority can be easily controlled by setting the thresholds differently. The units in the microgrid are autonomously operating using the bus voltage without communicating between units or to

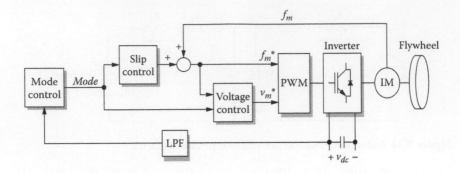

Figure 7.35 Block diagram of flywheel drive system.

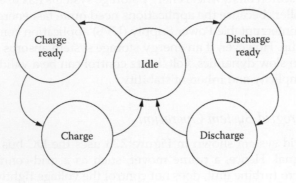

Figure 7.36 Controller state machine.

a central controller; hence, addition, replacement, or removal of the units can be done easily without any major change in the control configuration unlike the centrally controlled system.

7.5.4 Control of Flywheel Energy Storage System

The control block diagram of the flywheel drive system is shown in Figure 7.35. The controller consists of three parts—mode control, slip control, and voltage control—to generate the proper voltage and frequency for the flywheel induction motor or generator. The low-pass filter filters out the high-frequency voltage fluctuations. The mode control determines the operating mode based on the bus voltage. The operating modes of the proposed flywheel energy storage system, and the state machine of the mode control can be seen in Figure 7.36. Each mode has associated voltage levels predefined in the mode control, as shown in Figure 7.37.

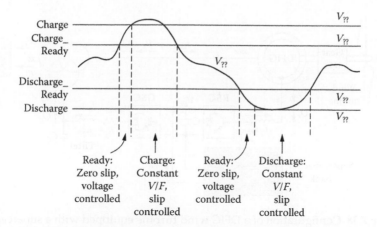

Figure 7.37 Bus voltage thresholds for control.

7.5.5 Stability Consideration

It is well known that DC power distribution systems can have stability issues due to the negative impedance of the connected converters, even if the individual subsystems are stable. The input impedance of the converter can be expressed as (7.13). The input impedance of the converters becomes negative when they are operating in constant power mode. where Δ denotes the deviation from the steady-state operating point values.

$$Z_i = \frac{\Delta v_{dc}}{\Delta i_{dc}} = -\frac{(v_{dc})^2}{P_0} \tag{7.13}$$

The power electronics converters can tightly control their output power as almost constant, and the negative impedance affects the DC bus stability adversely. It has also been suggested that the output impedances of the sources Z_0 should be smaller than input impedances of the loads Z_i for the overall stability of the DC bus.

$$|Z_0| << |Z_i| \tag{7.14}$$

Although this is not a direct issue for the proposed energy storage system because it does not operate in constant power mode, its effect on overall system stability needs to be considered. The proposed system controls the power with the slip and constant volt/hertz ratio, which keeps the torque proportional to the slip. When the flywheel energy storage

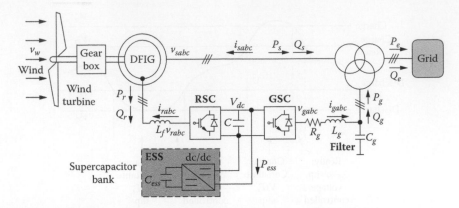

Figure 7.38 Configuration of a DFIG wind turbine equipped with a supercapacitor ESS connected to a power grid.

system charges the energy from the bus, the slip is proportional to the bus voltage. Hence, the power that the storage system takes from the bus is proportional to the bus voltage and the impedance of the system is always positive. On the other hand, the slip is inversely proportional to the change of the bus voltage, which lowers the overall source impedance because the DC current the storage system discharges increases as the bus voltage decreases. Therefore, the energy storage system can improve the overall stability of the system in either operating mode.

7.6   Constant Power Control of DFIG Wind Turbines with Supercapacitor Energy Storage

This section discusses a novel two-layer constant power control (CPC) scheme for a wind farm equipped with DFIG wind turbines, where each WTG is equipped with a supercapacitor energy storage system (ESS). The CPC consists of a high-layer wind farm supervisory controller (WFSC) and low-layer WTG controllers. The high-layer WFSC generates the active power references for the low-layer WTG controllers of each DFIG wind turbine according to the active power demand from the grid operator. The low-layer WTG controllers then regulate each DFIG wind turbine to generate the desired amount of active power, where the deviations between the available wind energy input and desired active power output are compensated by the ESS [67].

Figure 7.38 shows the basic configuration of a DFIG wind turbine equipped with a supercapacitor-based ESS. The low-speed wind turbine drives a high-speed DFIG through a gearbox. The DFIG is a wound-rotor induction machine. It is connected to the power grid at both stator and rotor terminals. The stator is directly connected to the grid, whereas the rotor is fed through a variable-frequency converter, which consists of a rotor-side converter (RSC) and a grid-side converter (GSC) connected back to back through a DC link and usually with a rating of a fraction (25–30%) of the DFIG nominal power. As a consequence, the WTG can operate with the rotational speed in a range of ±25–30% around the synchronous speed, and its active and reactive powers can be controlled independently. In this work, an ESS consisting of a supercapacitor bank and a two-quadrant DC-to-DC converter is connected to the DC link of the DFIG converters. The ESS serves as either a source or a sink of active power and therefore contributes to control the generated active power of the WTG. The value of the capacitance of the supercapacitor bank can be determined by

$$C_{ess} = \frac{2P_n T}{V_{SC}^2} \tag{7.15}$$

where C_{ess} is in farads, P_n is the rated power of the DFIG (watts), V_{SC} is the rated voltage of the supercapacitor bank (volts), and T is the desired time period (seconds) that the ESS can supply or store energy at the rated power (P_n) of the DFIG.

The use of an ESS in each WTG rather than a large single central ESS for the entire wind farm is based on two reasons. First, this arrangement has a high reliability because the failure of a single ESS unit does not affect the ESS units in other WTGs. Second, the use of an ESS in each WTG can reinforce the DC bus of the DFIG converters during transients, thereby enhancing the low-voltage ride-through capability of the WTG.

7.6.1 Control of Individual DFIG Wind Turbines

The control system of individual DFIG wind turbines generally consists of two parts: (1) the electrical control of the DFIG; and (2) the mechanical control of the wind turbine blade pitch angle and yaw system. Control of the DFIG is achieved by controlling the RSC, the GSC, and the ESS as shown in Figure 7.38. The control objective of the RSC is to regulate the stator-side active power P_s and reactive power Q_s independently. The control objective of the GSC is to maintain the DC link voltage V_{dc} constant and to regulate the reactive power Q_g that the GSC exchanges with the grid. The control objective of the ESS is to regulate the active power P_g that the GSC exchanges with the grid.

7.6.2 Control of the RSC

Figure 7.39 shows the overall vector control scheme of the RSC, in which the independent control of the stator active power P_s and reactive power

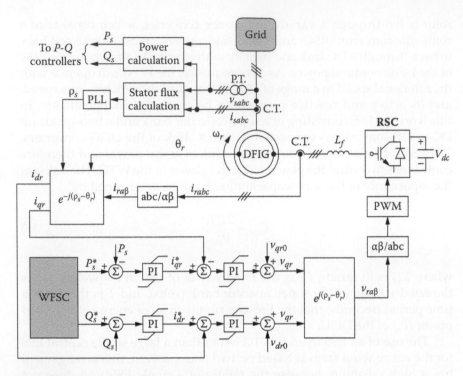

Figure 7.39 Overall vector control scheme of the RSC.

Q_s is achieved by means of rotor current regulation in a stator-flux-oriented synchronously rotating reference frame. Therefore, the overall RSC control scheme consists of two cascaded control loops. The outer control loop regulates the stator active and reactive powers independently, which generates the reference signals i^*_{dr} and i^*_{qr} of the d- and q-axis current components, respectively, for the inner-loop current regulation. The outputs of the two current controllers are compensated by the corresponding cross-coupling terms v_{dr0} and v_{qr0}, respectively, to form the total voltage signals v_{dr} and v_{qr}. They are then used by the PWM module to generate the gate control signals to drive the RSC. The reference signals of the outer-loop power controllers are generated by the high-layer WFSC.

7.6.3   Control of the GSC

Figure 7.40 shows the overall vector control scheme of the GSC, in which the control of the DC link voltage V_{dc} and the reactive power Q_g exchanged between the GSC and the grid is achieved by means of current regulation in a synchronously rotating reference frame. Again, the overall GSC control scheme consists of two cascaded control loops. The outer control

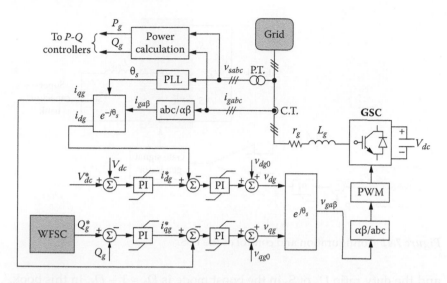

Figure 7.40 Overall vector control scheme of the GSC.

loop regulates the DC link voltage V_{dc} and the reactive power Q_g, respectively, which generates the reference signals i^*_{dg} and i^*_{qg} of the d- and q-axis current components, respectively, for the inner-loop current regulation. The outputs of the two current controllers are compensated by the corresponding cross-coupling terms v_{dg0} and v_{qg0}, respectively, to form the total voltage signals v_{dg} and v_{qg}. They are then used by the PWM module to generate the gate control signals to drive the GSC. The reference signal of the outer-loop reactive power controller is generated by the high-layer WFSC.

7.6.4 Configuration and Control of the ESS

Figure 7.41 shows the configuration and control of the ESS. The ESS consists of a supercapacitor bank and a two-quadrant DC-to-DC converter connected to the DC link of the DFIG. The DC-to-DC converter contains two IGBT switches S_1 and S_2. Their duty ratios are controlled to regulate the active power P_g that the GSC exchanges with the grid. In this configuration, the DC-to-DC converter can operate in two different modes (i.e., buck or boost mode) depending on the status of the two IGBT switches. If S_1 is open, the DC-to-DC converter operates in the boost mode; if S_2 is open, the DC-to-DC converter operates in the buck mode. The duty ratio D_1 of S_1 in the buck mode can be approximately expressed as

$$D_1 = \frac{V_{SC}}{V_{dc}} \qquad (7.16)$$

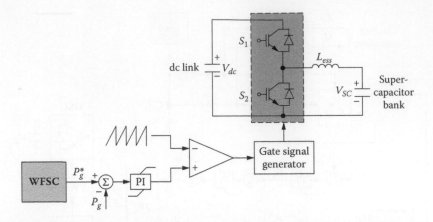

Figure 7.41 Configuration and control of the ESS.

and the duty ratio D_2 of S_2 in the boost mode is $D_2 = 1 - D_1$. In this book, the nominal DC voltage ratio $V_{SC/n}/V_{dc/n}$ is 0.5, where $V_{SC/n}$ and $V_{dc/n}$ are the nominal voltages of the supercapacitor bank and the DFIG DC link, respectively. Therefore, the nominal duty ratio $D_{1/n}$ of S_1 is 0.5.

The operating modes and duty ratios D_1 and D_2 of the DC-to-DC converter are controlled depending on the relationship between the active powers P_r of the RSC and P_g of the GSC. If P_r is greater than P_g, the converter is in buck mode and D_1 is controlled, such that the supercapacitor bank serves as a sink to absorb active power, which results in the increase of its voltage V_{SC}. On the contrary, if P_g is greater than P_r, the converter is in boost mode and D_2 is controlled, such that the supercapacitor bank serves as a source to supply active power, which results in the decrease of its voltage V_{SC}. Therefore, by controlling the operating modes and duty ratios of the DC-to-DC converter, the ESS serves as either a source or a sink of active power to control the generated active power of the WTG. In Figure 7.41, the reference signal P^*_g is generated by the high-layer WFSC.

7.6.5   Wind Turbine Blade Pitch Control

Figure 7.42 shows the blade pitch control for the wind turbine, where ω_r and P_e $(= P_s + P_g)$ are the rotating speed and output active power of the DFIG, respectively. When the wind speed is below the rated value and the WTG is required to generate the maximum power, ω_r and P_e are set at their reference values, and the blade pitch control is deactivated. When the wind speed is below the rated value but the WTG is required to generate a constant power less than the maximum power, the active power controller may be activated, where the reference signal P^*_e is generated by the high-layer WFSC and P_e takes the actual measured value. The active power

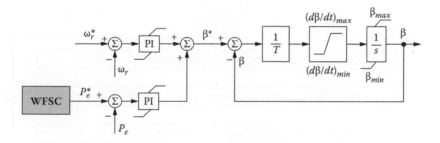

Figure 7.42 Blade pitch control for the wind turbine.

controller adjusts the blade pitch angle to reduce the mechanical power that the turbine extracts from wind. This reduces the imbalance between the turbine mechanical power and the DFIG output active power, thereby reducing the mechanical stress in the WTG and stabilizing the WTG system. Finally, when the wind speed increases above the rated value, both ω_r and P_e take the actual measured values, and both the speed and active power controllers are activated to adjust the blade pitch angle.

7.6.6   Wind Farm Supervisory Control

The objective of the WFSC is to generate the reference signals for the outer-loop power controllers of the RSC and GSC, the controller of the DC-to-DC converter, and the blade pitch controller of each WTG, according to the power demand from or the generation commitment to the grid operator. The implementation of the WFSC is described by the flow chart in Figure 7.43, where P_d is the active power demand from or the generation commitment to the grid operator; v_{wi} and V_{essi} are the wind speed in meters per second and the voltage of the supercapacitor bank measured from WTG i ($i = 1, \ldots, N$), respectively; and N is the number of WTGs in the wind farm.

7.7   Output Power Leveling of Wind Generator Systems by Pitch Angle Control

In medium- to large-size wind turbine generator systems, control of the pitch angle is a usual method for output power control in previously rated wind speed. Several control methods for controlling the pitch angle have been reported so far, such as the back stepping method and the feed-forward method. However, the variation in parameters and the effect of wind shear for windmill have not been considered in these methods. Considering this, the pitch angle control using minimum variance control and generalized predictive control (GPC) was reported. The aforementioned methods have a fixed pitch angle at $10°$ in below-rated wind

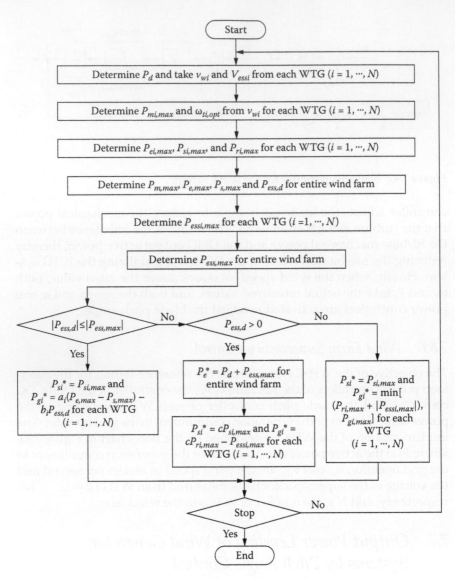

Figure 7.43 Flow chart of implementation of the WFSC.

speed; however, an actual wind speed distribution has much higher probability in below-rated wind speed. Thus, if many WTGs using squirrel-cage induction generators are interconnected to the power system, output power fluctuation is supplied to the power system. The VS WTG occurs in similar situations because the VS WTG in below-rated wind speed is based on the maximum energy capture strategy that is corresponding to wind speed variation. But the leveling of output power has a problem as

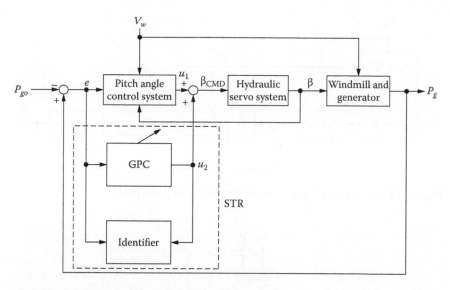

Figure 7.44 Pitch angle control system using GPC.

the output power reduces in below-rated wind speed. It is evident that a large-scale wind farm could be increased in the near future. Thus, in all operating regions, the output power fluctuation control of stand-alone WTG becomes important [60].

In this context, output power leveling of WTG for all operating regions by pitch angle control is discussed. The method presents a control strategy based on the average wind speed and standard deviation of wind speed and the pitch angle control, using GPC in all operating regions for WTG. The output power command is determined by approximate equation for windmill output using average wind speed. The output power of WTG for all operating regions is leveled by GPC, which is based on output power command. Thus the WTG is capable of providing stability operation for rapid change in operating point. And with this method it is possible to level output power of WTG for all operating regions by pitch angle control. Moreover, this pitch angle control can be used regardless of the kind of generators such as a permanent magnet synchronous generator (PMSG), a synchronous generator (SG), and DFIG.

The pitch angle control system using a GPC is shown in Figure 7.44, where $Pgo(k)$ is output power command, $Pg(k)$ is output power, $e(k)$ is output power error of generator, $u2(k)$ is control input of self-tuning regulator (STR), and k is number of sampling. The windmill and generator system is shown in Figure 7.45. Figure 7.46 shows the pitch angle control system that resolves pitch angle command β_{CMD}, where output power error e is used as input into potential difference (PD) controller. The hydraulic servo system is shown in Figure 7.47. The system actually has nonlinear characteristics,

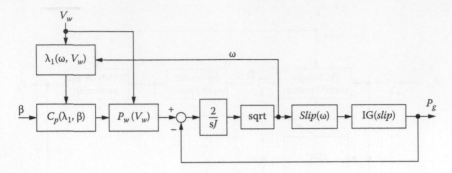

Figure 7.45 System configuration of windmill and generator.

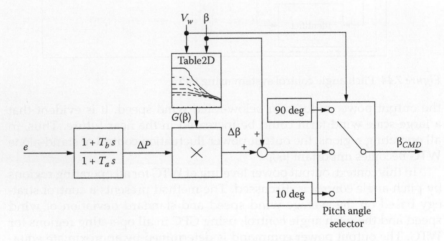

Figure 7.46 Pitch angle control system.

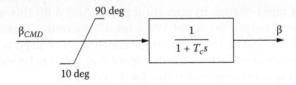

Figure 7.47 Hydraulic servo system.

but it is able to make a first-order lag system. The pitch angle command β_{CMD} is limited by a limiter at the range of 10–90°.

The conventional method for the pitch angle law is fixed at more than cut-in wind speed and less than rated wind speed so that the output power for WTG is proportional to the fluctuation of wind speed at

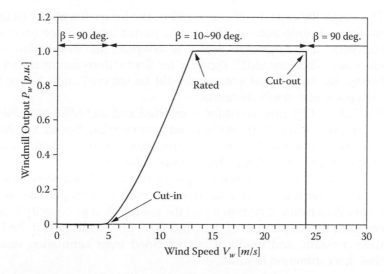

Figure 7.48 Pitch angle control law for all operating regions.

more than cut-in wind speed and less than rated wind speed. Thus, to achieve output power leveling of WTG for all operating regions by pitch angle control, the pitch angle control law has been extended as shown in Figure 7.48, whereas the fixed rated output power command has been converted to variable output power command.

7.8 Chapter Summary

This chapter deals with fluctuation minimization of power, frequency, and voltage by energy storage devices, especially using SMES systems. Various energy storage systems are discussed. A comparison among energy storage systems is also done. A multimachine power system consisting of hydraulic generators, diesel generators, and a fixed-speed wind generator is considered. The performance of the SMES is evaluated by considering their different load demands and energy capacities. Also, an attempt is made to evaluate the power and energy ratings of SMES systems. The required SMES power rating is analyzed using the standard deviation of frequency fluctuations with respect to the LPF time constant. From the simulation results, the following points are noteworthy.

1. By using the SMES system, the system frequency fluctuations are successfully minimized, and the frequency is maintained almost at the level of the nominal value of 60 Hz for different load demands.
2. The SMES can successfully minimize voltage fluctuations and can smooth the transmission line power.

3. The larger the capacity of the SMES is, the better the ability of fluctuations minimization of frequency, power, and voltage becomes. However, if a large-size SMES is adopted, its installation cost increases. Thus, the SMES capacity for fluctuations minimization of power, frequency, and voltage should be selected considering the viewpoint of cost-effectiveness.

4. If a 120 sec LPF time constant is selected and an SMES unit with a 55% power rating of that of the wind generator is adopted, a suitable reference value for the compensating power can be obtained and then sufficient smoothing effect can be achieved.

5. The required energy storage capacity of the SMES system is estimated under wind power fluctuations with various periodic wave forms. As a result, it is shown that the energy storage capacity can be obtained by the product of the wind farm power rating and the LPF time constant, and finally it is confirmed from simulation results that this estimation is valid.

Furthermore, this chapter discusses the output power leveling of wind generator systems by pitch angle control, power quality improvement by flywheel energy storage system, and constant power control of DFIG wind turbines with supercapacitor energy storage.

References

1. P. F. Ribeiro, B. K. Johnson, M. L. Crow, A. Arsoy, and Y. Liu, "Energy storage systems for advanced power applications," *Proceedings of the IEEE*, vol. 89, no. 12, pp. 1744–1756, December 2001.

2. M. H. Ali, J. Tamura, and B. Wu, "SMES strategy to minimize frequency fluctuations of wind generator system," *Proceedings of the 34th Annual Conference of the IEEE Industrial Electronics Society (IECON 2008)*, November 10–13, 2008, Orlando, FL, pp. 3382–3387.

3. T. Asao, R. Takahashi, T. Murata, J. Tamura, M. Kubo, Y. Matsumura, et al., "Evaluation method of power rating and energy capacity of superconducting magnetic energy storage system for output smoothing control of wind farm," *Proceedings of the 2008 International Conference on Electrical Machines*, pp. 1–6, 2008.

4. T. Asao, R. Takahashi, T. Murata, J. Tamura, M. Kubo, A. Kuwayama, et al., "Smoothing control of wind power generator output by superconducting magnetic energy storage system," *Proceedings of the 2007 International Conference on Electrical Machines*, pp. 1–6, 2007.

5. H. J. Boenig and J. F. Hauer, "Commissioning tests of the Bonneville Power Administration 30 MJ superconducting magnetic energy storage unit," *IEEE Trans. Power Apparatus and Systems*, vol. PAS-104, no. 2, pp. 302–309, February 1985.

6. S. C. Tripathy, M. Kalantar, and R. Balasubramanian, "Dynamics and stability of wind and diesel turbine generators with superconducting magnetic energy storage on an isolated power system," *IEEE Trans. Energy Conversion*, vol. 6, no. 4, pp. 579–585, December 1991.

7. M. D. Mufti, R. Balasubramanian, S. C. Tripathy, and S. A. Lone, "Modelling and control of weak power systems supplied from diesel and wind," *International Journal of Power and Energy Systems*, vol. 23, no. 1, pp. 24–36, 2003.

8. S. Nomura, Y. Ohata, T. Hagita, H. Tsutsui, S. Tsuji-Iio, and R. Shimada, "Wind farms linked by SMES systems," *IEEE Trans. Applied Superconductivity*, vol. 15, no. 2, pp. 1951–1954, June 2005.

9. F. Zhou, G. Joos, C. Abbey, L. Jiao, and B. T. Ooi, "Use of large capacity SMES to improve the power quality and stability of wind farms," *IEEE Power Engineering Society General Meeting, 2004*, vol. 2, pp. 2025–2030, June 2004.

10. M. H. Ali, M. Park, L.-K. Yu, T. Murata, and J. Tamura, "Improvement of wind generator stability by fuzzy logic-controlled SMES," *Proceeding of the International Conference on Electrical Machines and Systems 2007 (ICEMS 2007)*, pp. 1753–1758, October 8–11, 2007.

11. M. H. Ali, T. Murata, and J. Tamura, "Minimization of fluctuations of line power and terminal voltage of wind generator by fuzzy logic-controlled SMES," *International Review of Electrical Engineering (IREE)*, vol. 1, no. 4, pp. 559–566, October 2006.

12. T. Kinjo, T. Senjyu, N. Urasaki, and H. Fujita, "Terminal-voltage and output-power regulation of wind-turbine generator by series and parallel compensation using SMES," *IEE Proc.-Gener. Transm. Distrib.*, vol. 153, no. 3, pp. 276–282, May 2006.

13. J. J. Skiles, R. L. Kustom, K.-P. Ko, V. Wong, K.-S. Ko, F. Vong, et al., "Performance of a power conversion system for superconducting magnetic energy storage (SMES)," *IEEE Trans. Power Systems*, vol. 11, no. 4, pp. 1718–1723, November 1996.

14. S. C. Tripathy and K. P. Juengst, "Sampled data automatic generation control with superconducting magnetic energy storage in power systems," *IEEE Trans. Energy Conversion*, vol. 12, no. 2, pp. 187–192, June 1997.

15. IEEE Task Force on Benchmark Models for Digital Simulation of FACTS and Custom–Power Controllers, T&D Committee, "Detailed modeling of superconducting magnetic energy storage (SMES) system," *IEEE Trans. Power Delivery*, vol. 21, no. 2, pp. 699–710, April 2006.

16. *PSCAD/EMTDC manual*, Manitoba HVDC Research Center, 1994.

17. S. Roy, O. P. Malik, and G. S. Hope, "A k-step predictive scheme for speed control of diesel driven power plants," *IEEE Trans. Industry Applications*, vol. 29, no. 2, pp. 389–396, March–April 1993.

18. G. S. Stavrakakis and G. N. Kariniotakis, "A general simulation algorithm for the accurate assessment of isolated diesel–wind turbines systems interaction. Part 1: A general multimachine power system model," *IEEE Trans. Energy Conversion*, vol. 10, no. 3, pp. 577–583, September 1995.

19. G. N. Kariniotakis and G. S. Stavrakakis, "A general simulation algorithm for the accurate assessment of isolated diesel–wind turbines systems interaction. Part 2: Implementation of the algorithm and case-studies with induction generators," *IEEE Trans. Energy Conversion*, vol. 10, no. 3, pp. 584–590, September 1995.

20. S. Roy, O. P. Malik, and G. S. Hope, "An adaptive control scheme for speed control of diesel driven power-plants," *IEEE Trans. Energy Conversion*, vol. 6, no. 4, pp. 605–611, December 1991.

22. S. Heier, *Grid integration of wind energy conversion system*, John Wiley & Sons, 1998.

23. P. M. Anderson and A. Bose, "Stability simulation of wind turbine systems," *IEEE Trans. Power Apparatus and Systems*, vol. PAS-102, no. 12, pp. 3791–3795, December 1983.

24. R. Dugan and T. McDermott, "Distributed generation," *IEEE Industry Applications Magazine*, vol. 8, no. 2, pp. 19–25, March–April 2002.

24. P. Chiradeja and R. Ramakumar, "An approach to quantify the technical benefits of distributed generation," *IEEE Transactions on Energy Conversion*, vol. 19, no. 4, pp. 764–773, December 2004.

25. F. Blaabjerg, R. Teodorescu, M. Liserre, and A. Timbus, "Overview of control and grid synchronization for distributed power generation systems," *IEEE Transactions on Industrial Electronics*, vol. 53, no. 5, pp. 1398–1407, October 2006.

26. H. Nikkhajoei and R. Lasseter, "Distributed generation interface to the CERTS microgrid," *IEEE Transactions on Power Delivery*, vol. 24, no. 3, pp. 1598–1608, July 2009.

27. F. Katiraei, R. Iravani, N. Hatziargyriou, and A. Dimeas, "Microgrids management," *IEEE Power & Energy Magazine*, pp. 54–65, May–June 2008.

28. D. Salomonsson and A. Sannino, "Low-voltage DC distribution system for commercial power systems with sensitive electronic loads," *IEEE Transactions on Power Delivery*, vol. 22, no. 3, pp. 1620–1627, July 2007.

29. P. Tsao, M. Senesky, and S. R. Sanders, "An integrated flywheel energy storage system with homopolar inductor motor/generator and high frequency drive," *IEEE Transactions on Industry Applications*, vol. 39, no. 6, pp. 1710–1725, November–December 2003.

30. J.-D. Park, C. Kalev, and H. Hofmann, "Control of high-speed solid rotor synchronous reluctance motor/generator for flywheel-based uninterruptible power supplies," *IEEE Transactions on Industrial Electronics*, vol. 55, no. 8, pp. 3038–3046, February 2008.

31. H. Akagi and H. Sato, "Control and performance of a doubly-fed induction machine intended for a flywheel energy storage system," *IEEE Transactions on Power Electronics*, vol. 17, no. 1, pp. 109–116, August 2002.

32. R. Cardenas, R. Pena, G. Asher, and J. Clare, "Control strategies for enhanced power smoothing in wind energy systems using a flywheel driven by a vector-controlled induction machine," *IEEE Transactions on Industrial Electronics*, vol. 48, no. 3, pp. 625–635, June 2001.

33. X.-D. Sun, K.-H. Koh, B.-G. Yu, and M. Matsui, "Fuzzy-logic-based V/f control of an induction motor for a DC grid power-leveling system using flywheel energy storage equipment," *IEEE Transactions on Industrial Electronics*, vol. 56, no. 8, pp. 3161–3168, May 2009.

34. R. Cardenas, R. Pena, G. Asher, J. Clare, and R. Blasco-Gimenez, "Control strategies for power smoothing using a flywheel driven by a sensorless vector-controlled induction machine operating in a wide speed range," *IEEE Transactions on Industrial Electronics*, vol. 51, no. 3, pp. 603–614, June 2004.

35. X. Feng, J. Liu, and F. Lee, "Impedance specifications for stable DC distributed power systems," *IEEE Transactions on Power Electronics*, vol. 17, no. 2, pp. 157–162, March 2002.

36. H. Mosskull, J. Galic, and B. Wahlberg, "Stabilization of induction motor drives with poorly damped input filters," *IEEE Transactions on Industrial Electronics*, vol. 54, no. 5, pp. 2724–2734, October 2007.

37. R. Piwko, D. Osborn, R. Gramlich, G. Jordan, D. Hawkins, and K. Porter, "Wind energy delivery issues: Transmission planning and competitive electricity market operation," *IEEE Power Energy Mag.*, vol. 3, no. 6, pp. 47–56, November–December 2005.

38. L. Landberg, G. Giebel, H. A. Nielsen, T. Nielsen, and H. Madsen, "Short term prediction—An overview," *Wind Energy*, vol. 6, no. 3, pp. 273–280, July–September 2003.

39. J. P. Barton and D. G. Infield, "Energy storage and its use with intermittent renewable energy," *IEEE Trans. Energy Conversion*, vol. 19, no. 2, pp. 441–448, June 2004.

40. C. Abbey and G. Joos, "Supercapacitor energy storage for wind energy applications," *IEEE Trans. Ind. Appl.*, vol. 43, no. 3, pp. 769–776, May–June 2007.

41. B. S. Borowy and Z. M. Salameh, "Dynamic response of a stand-alone wind energy conversion system with battery energy storage to wind gust," *IEEE Trans. Energy Conversion*, vol. 12, no. 1, pp. 73–78, March 1997.

42. M.-S. Lu, C.-L. Chang, W.-J. Lee, and L. Wang, "Combining the wind power generation system with energy storage equipments," *IEEE Trans. Ind. Appl.*, vol. 45, no. 6, pp. 2109–2115, November–December 2009.

43. W. Qiao, W. Zhou, J. M. Aller, and R. G. Harley, "Wind speed estimation based sensorless output maximization control for a wind turbine driving a DFIG," *IEEE Trans. Power Electron.*, vol. 23, no. 3, pp. 1156–1169, May 2008.

44. W. Qiao, G. K. Venayagamoorthy, and R. G. Harley, "Real-time implementation of a STATCOM on a wind farm equipped with doubly fed induction generators," *IEEE Trans. Ind. Appl.*, vol. 45, no. 1, pp. 98–107, January–February 2009.

45. S. Vazquez, S. M. Lukic, E. Galvan, L. G. Franquelo, and J. M. Carrasco, "Energy storage systems for transport and grid applications," *IEEE Trans. Industrial Electronics*, vol. 57, no. 12, pp. 3881–3895, December 2010.

46. E. Manla, A. Nasiri, C. H. Rentel, and M. Hughes, "Modeling of zinc bromide energy storage for vehicular applications," *IEEE Trans. Ind. Electron.*, vol. 57, no. 2, pp. 624–632, February 2010.

47. D. P. Scamman, G. W. Reade, and E. P. L. Roberts, "Numerical modeling of a bromide-polysulphide redox flow battery. Part 1: Modelling approach and validation for a pilot-scale system," *J. Power Sources*, vol. 189, no. 2, pp. 1220–1230, April 2009.

48. D. P. Scamman, G. W. Reade, and E. P. L. Roberts, "Numerical modeling of a bromide-polysulphide redox flow battery. Part 2: Evaluation of a utility-scale system," *J. Power Sources*, vol. 189, no. 2, pp. 1231–1239, April 2009.

49. S. Lemofouet and A. Rufer, "A hybrid energy storage system based on compressed air and supercapacitors with maximum efficiency point tracking (MEPT)," *IEEE Trans. Ind. Electron.*, vol. 53, no. 4, pp. 1105–1115, June 2006.

50. D. J. Swider, "Compressed air energy storage in an electricity system with significant wind power generation," *IEEE Trans. Energy Conversion*, vol. 22, no. 1, pp. 95–102, March 2007.

51. M. B. Camara, H. Gualous, F. Gustin, and A. Berthon, "Design and new control of DC/DC converters to share energy between supercapacitors and batteries in hybrid vehicles," *IEEE Trans. Veh. Technol.*, vol. 57, no. 5, pp. 2721–2735, September 2008.

52. H. Yoo, S. K. Sul, Y. Park, and J. Jeong, "System integration and power flow management for a series hybrid electric vehicle using supercapacitors and batteries," *IEEE Trans. Ind. Appl.*, vol. 44, no. 1, pp. 108–114, January–February 2008.

53. L. Shuai, K. A. Corzine, and M. Ferdowsi, "A new battery/ultracapacitor energy storage system design and its motor drive integration for hybrid electric vehicles," *IEEE Trans. Veh. Technol.*, vol. 56, no. 4, pp. 1516–1523, July 2007.

54. W. Henson, "Optimal battery/ultracapacitor storage combination," *J. Power Sources*, vol. 179, no. 1, pp. 417–423, April 2008.

55. K. Jin, X. Ruan, M. Yang, and M. Xu, "A hybrid fuel cell power system," *IEEE Trans. Ind. Electron.*, vol. 56, no. 4, pp. 1212–1222, April 2009.

56. M. H. Todorovic, L. Palma, and P. N. Enjeti, "Design of a wide input range DC–DC converter with a robust power control scheme suitable for fuel cell power conversion," *IEEE Trans. Ind. Electron.*, vol. 55, no. 3, pp. 1247–1255, March 2008.

57. M. Ortuzar, J. Moreno, and J. Dixon, "Ultracapacitor-based auxiliary energy system for an electric vehicle: Implementation and evaluation," *IEEE Trans. Ind. Electron.*, vol. 54, no. 4, pp. 2147–2156, August 2007.

58. O. Briat, J. M. Vinassa, W. Lajnef, S. Azzopardi, and E. Woirgard, "Principle, design and experimental validation of a flywheel-battery hybrid source for heavy-duty electric vehicles," *IET Elect. Power Appl.*, vol. 1, no. 5, pp. 665–674, September 2007.

59. T. Ise, M. Kita, and A. Taguchi, "A hybrid energy storage with a SMES and secondary battery," *IEEE Trans. Appl. Supercond.*, vol. 15, no. 2, pp. 1915–1918, June 2005.

60. T. Senjyu, R. Sakamoto, N. Urasaki, T. Funabashi, H. Fujita, and H. Sekine, "Output power leveling of wind turbine generator for all operating regions by pitch angle control," *IEEE Trans. Energy Conversion*, vol. 21, no. 2, pp. 467–475, June 2006.

61. E. Muljadi and C. P. Butterfield, "Pitch-controlled variable-speed wind turbine generation," *IEEE Trans. Ind. Appl.*, vol. 37, no. 1, pp. 240–246, January–Februar 2001.

62. J. L. Rodriguez-Amenedo, S. Arnalte, and J. C. Burgos, "Automatic generation control of a wind farm with variable speed wind turbines," *IEEE Trans. Energy Conversion*, vol. 17, no. 2, pp. 279–284, June 2002.

63. A. Miller, E. Muljadi, and D. S. Zinger, "A variable speed wind turbine power control," *IEEE Trans. Energy Conversion*, vol. 12, no. 2, pp. 181–186, June 1997.

64. P. Ledesma and J. Usaola, "Doubly fed induction generator model for transient stability analysis," *IEEE Trans. Energy Conversion*, vol. 20, no. 2, pp. 388–397, June 2005.

65. C. Carrillo, A. E. Feijóo, J. Cidrás, and J. González, "Power fluctuations in an isolated wind plant," *IEEE Trans. Energy Convers.*, vol. 19, no. 1, pp. 217–221, March 2004.

66. Jae-Do Park, "Simple flywheel energy storage using squirrel-cage induction machine for DC bus microgrid systems," Proceedings of the 36th Annual Conference on IEEE Industrial Electronics Society, IECON 2010, 7-10 Nov. 2010, pp. 3040-3045, Glendale, AZ, USA.

67. L. Qu and W. Qiao, "Constant power control of DFIG wind turbines with supercapacitor energy storage," *IEEE Trans. Industry Applications*, vol. 47, no. 1, pp. 359–367, January/February 2011.

chapter 8

Solutions for Transient Stability Issues of Fixed-Speed Wind Generator Systems

8.1 Introduction

Induction machines are mostly used as wind generators. However, induction machines have stability problems, similar to the transient stability of synchronous machines [1–6]. Therefore, it is important to analyze the transient stability of power systems including wind power stations. The research on the selection of a suitable device for the stabilization of fixed-speed wind generators is a matter of interest. In literature, the static synchronous compensator (STATCOM) is reported to stabilize the fixed-speed wind generator [7–9]. The braking resistor (BR) has been recognized and used as a cost-effective measure for transient stability control of synchronous generators for a long time. According to some recent reports, BRs can be used for wind generator stabilization as well [10–16]. A controller of the blade pitch angle of a windmill is, in general, equipped with wind turbines. Although the main purpose of the pitch controller is to maintain output power of a wind generator at the rated level when wind speed is above the rated speed, it can enhance the transient stability of the wind generator by controlling the rotor speed [17–19].

Superconducting magnetic energy storage (SMES) is a large superconducting coil capable of storing electric energy in the magnetic field generated by direct current (DC) current flowing through it. The real as well as reactive power can be absorbed (charging) by or released (discharging) from the SMES coil according to system power requirements. Since the successful commissioning test of the BPA 30 MJ unit, SMES systems have received much attention in power system applications, such as diurnal load demand leveling, frequency control, automatic generation control, and uninterruptible power supplies. Recently, research has been conducted to investigate the effectiveness of SMES as a tool for the stabilization of grid-connected wind generator systems [20–25]. Superconducting fault current limiters (SFCLs) can suppress short-circuit currents using the unique quench characteristics of superconductors. In the event of a fault, the superconductor undergoes a transition into its normal state (i.e., quenching). After quenching, the current is commutated to a shunt resistance and is then limited rapidly [26–30]. Using this excellent performance of superconductors, the SFCL has been demonstrated as a tool to stabilize the variable-speed wind generator system [31–32].

This chapter discusses the aforementioned stabilization methods of wind generator systems in detail and also compares them. The analysis is performed in terms not only of transient stability enhancement but also of controller complexity and cost. One of the salient features of this chapter is that the transient stability analysis of wind generator systems is carried out considering the unsuccessful reclosing of circuit breakers. The effectiveness of each stabilization method is demonstrated through simulation results. It is hoped that this chapter would help readers understand the relative effectiveness of these stabilization methods and would provide a guideline for selecting a suitable technique for the stabilization of wind energy systems.

8.2   Model System

To analyze transient stability, the model system shown in Figure 8.1 was considered [1]. The model system consists of one synchronous generator (100 MVA, SG) and one wind turbine generator (50 MVA induction generator, IG), which deliver power to an infinite bus through a transmission line with two circuits. Though a wind power station is composed of many generators practically, it is considered to be composed of a single generator with the total power capacity in this chapter. There is a local transmission line with one circuit between the main transmission line and a transformer at the wind power station. A capacitor C is connected to the

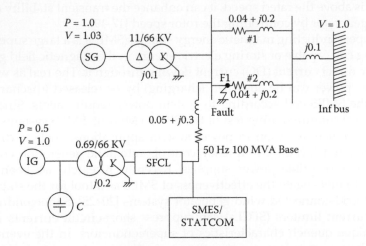

Figure 8.1 Power system model.

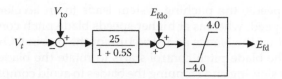

Figure 8.2 AVR model.

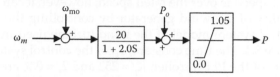

Figure 8.3 GOV model.

Table 8.1 Generator Parameters

	SG		IG
MVA	100	MVA	50
r_a (pu)	0.003	r_1 (pu)	0.01
x_a (pu)	0.13	x_1 (pu)	0.18
X_d (pu)	1.2	X_{mu} (pu)	10.0
X_q (pu)	0.7	r_2 (pu)	0.015
X'_d (pu)	0.3	x_2 (pu)	0.12
X'_q (pu)	0.22	H (sec)	0.75
X''_d (pu)	0.22		
X''_q (pu)	0.25		
T'do (sec)	5.0		
T"do (sec)	0.04		
T"qo (sec)	0.05		

terminal of the wind generator to compensate the reactive power demand for the induction generator at the steady-state. The value of C has been chosen so that the power factor of the wind power station becomes unity when it is generating the rated power (P = 0.5, V = 1.0). The automatic voltage regulator (AVR) and governor (GOV) control system models shown in Figures 8.2 and 8.3, respectively, for the synchronous generator are considered. Table 8.1 shows the synchronous generator parameters as well as induction generator parameters used for the simulation analysis [33–34].

8.3 Pitch Control Method

Blade pitch control is the system that monitors and adjusts the inclination angle of the blades and thus controls the rotation speed of the blades. At lower wind speeds, the pitching system leads to an acceleration of the hub rotation speed, whereas at higher speeds blade pitch control reduces the wind load on the blades and structure of the turbine. Over a certain wind speed the blade pitch control starts to rotate the blades out of the wind, thereby slowing and stopping the blades to avoid complete damage. Thus, the main purpose of using a pitch controller with a wind turbine is to maintain a constant output power at the terminal of the generator when the wind speed is over the rated speed; however, it can enhance the transient stability of the wind generator by controlling the rotor speed. The pitch control system model of the wind turbine used in this book is shown in Figure 8.4. The time constant, T_d, of the control system is 3.0 sec. The parameters of the PI controller, $K_p = 252$ and $T_i = 0.3$, are determined by trial and error to obtain the best system performance.

8.4   Superconducting Magnetic Energy Storage Method

An SMES device is a DC current device that stores energy in the magnetic field. The DC current flowing through a superconducting wire in a large magnet creates the magnetic field. Figure 8.5 shows the basic configuration

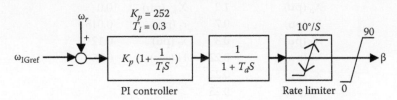

Figure 8.4 Pitch control system model.

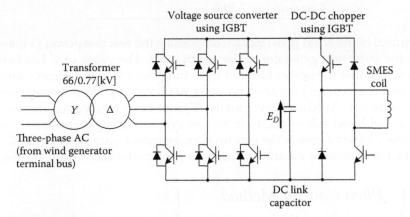

Figure 8.5 Basic configuration of SMES system.

of the proposed SMES unit, which consists of a Wye-Delta 66 KV/0.77 KV transformer, a six-pulse pulse width modulation (PWM) rectifier/inverter (50 MVA) using an insulated gate bipolar transistor (IGBT), a two-quadrant DC-to-DC chopper using an IGBT, and a superconducting coil or inductor of 0.24 H. The PWM converter and the DC-to-DC chopper are linked by a DC link capacitor of 60 mF [36]. The detailed explanation of the voltage source converter (VSC) and the two-quadrant DC-to-DC chopper is described next.

For an SMES system, the inductively stored energy (E in Joule) and the rated power (P in watts) are commonly the given specifications for SMES devices and can be expressed as follows:

$$E = \frac{1}{2} L_{sm} I_{sm}^{2}$$

$$P = \frac{dE}{dt} = L_{sm} I_{sm} \frac{dI_{sm}}{dt} = V_{sm} I_{sm}$$

(8.1)

where L_{sm} is the inductance of the coil, I_{sm} is the DC current flowing through the coil, and V_{sm} is the voltage across the coil. The SMES unit is located at the wind generator terminal bus. It is considered that the SMES has the rating of 50 MW and 0.05 MWh.

8.4.1   PWM Voltage Source Converter

The PWM VSC provides a power electronic interface between the alternating current (AC) power system and the superconducting coil. In the PWM generator, the sinusoidal reference signal is phase modulated by means of the phase angle, a, of the VSC output AC voltage. The phase angle, a, is determined from the outputs of the proportional-integral (PI) controllers as shown in Figure 8.6, where ΔV_{cap} and ΔV_{IG} indicate the capacitor voltage deviation and terminal voltage deviation of the induction generator, respectively. The PI controller parameters are determined by trial and error to obtain the best system performance. In this book, the amplitude modulation index of the sinusoidal reference signal is chosen as 1.0. The modulated sinusoidal reference signal is compared with the triangular carrier signal to generate the gate signals for the IGBTs. The frequency of the triangular carrier signal is chosen as 450 Hz. The DC voltage across the capacitor is 1,000 volts, which is kept constant throughout by the six-pulse PWM converter.

8.4.2   Two-Quadrant DC-to-DC Chopper

The superconducting coil is charged or discharged by adjusting the average (i.e., DC) voltage across the coil to be positive or negative values by

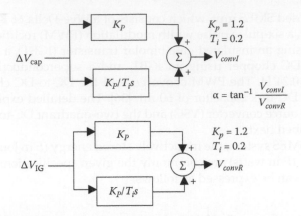

Figure 8.6 Block diagrams of PI controller to determine a.

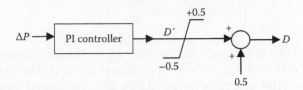

Figure 8.7 Control of chopper duty cycle.

means of the DC-to-DC chopper duty cycle, D, controlled by a conventional PI controller as shown in Figure 8.7, where DP indicates the real power deviation of the induction generator. When the duty cycle is larger than 0.5 or less than 0.5, the coil is either charging or discharging, respectively. When the unit is on standby, the coil current is kept constant, independent of the storage level, by adjusting the chopper duty cycle to 50%, resulting in the net voltage across the superconducting winding to be zero. To generate the gate signals for the IGBTs of the chopper, the PWM reference signal is compared with the saw tooth carrier signal. The frequency of the saw tooth carrier signal for the chopper is chosen as 100 Hz.

8.5 *Static Synchronous Compensator (STATCOM) Method*

A static synchronous compensator (STATCOM) is a second-generation flexible AC transmission system controller based on a self-commutating solid-state voltage source inverter. It is a shunt-connected reactive compensation equipment that is capable of generating or absorbing reactive power whose output can be varied to maintain control of specific parameters of the electric power system. As can be seen from Figure 8.5, excluding the DC-to-DC chopper and SMES coil the remaining components represent the basic two-level STATCOM, which is used in this book.

The STATCOM provides operating characteristics similar to a rotating synchronous compensator without the mechanical inertia; since the STATCOM employs solid-state power switching devices, it provides rapid controllability of the three phase voltages, both in magnitude and phase angle. The STATCOM basically consists of a step-down transformer with a leakage reactance, a three-phase GTO or IGBT voltage source inverter (VSI), and a DC capacitor. The AC voltage difference across the leakage reactance produces reactive power exchange between the STATCOM and the power system, such that the AC voltage at the bus bar can be regulated to improve the voltage profile of the power system, which is the primary duty of the STATCOM. However, for instance, a secondary damping function can be added into the STATCOM for enhancing power system oscillation stability.

The principle of STATCOM operation is as follows. The VSI generates a controllable AC voltage source behind the leakage reactance. This voltage is compared with the AC bus voltage system; when the AC bus voltage magnitude is above that of the VSI voltage magnitude, the AC system sees the STATCOM as an inductance connected to its terminals. Otherwise, if the VSI voltage magnitude is above that of the AC bus voltage magnitude, the AC system sees the STATCOM as a capacitance connected to its terminals. If the voltage magnitudes are equal, the reactive power exchange is zero. If the STATCOM has a DC source or energy storage device on its DC side, it can supply real power to the power system. This can be achieved by adjusting the phase angle of the STATCOM terminals and the phase angle of the AC power system. When the phase angle of the AC power system leads the VSI phase angle, the STATCOM absorbs real power from the AC system; if the phase angle of the AC power system lags the VSI phase angle, the STATCOM supplies real power to AC system.

Typical applications of STATCOM are effective voltage regulation and control, reduction of temporary overvoltages, improvement of steady-state power transfer capacity, improvement of transient stability margin, damping of power system oscillations, damping of subsynchronous power system oscillations, flicker control, power quality improvement, and distribution system applications.

The VSC or VSI is the building block of a STATCOM and other flexible AC transmission systems (FACTS) devices. A very simple inverter produces a square voltage waveform as it switches the direct voltage source on and off. The basic objective of a VSI is to produce a sinusoidal AC voltage with minimal harmonic distortion from a DC voltage. Three basic techniques are used for reducing harmonics in the converter output voltage: (1) harmonic neutralization using magnetic coupling (multipulse converter configurations); (2) harmonic reduction using multilevel converter configurations; and (3) the PWM.

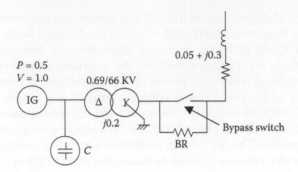

Figure 8.8 Location of the BR.

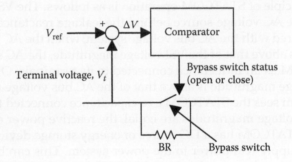

Figure 8.9 BR control method.

8.6   Braking Resistor Method

The BR concept aims to contribute directly to the balance of active power during a fault. It can be done by dynamically inserting a resistor in the generation circuit, increasing the voltage at the terminals of the generator, and thereby mitigating the destabilizing depression of electrical torque and power during the fault period. This book uses a series dynamic braking resistor of 1.0 pu (43.56 ohm) value for wind generator stabilization. The location of the BR in the power system model of Figure 8.1 is shown in Figure 8.8.

The BR would operate with its parallel switch closed under normal conditions, bypassing the braking resistor. Voltage depression below a selected set point would lead to near instantaneous tripping of the switch. Current would then flow through the inserted resistor, dissipating power. The braking resistor would remain in the circuit as long as the terminal voltage of the wind generator is below a threshold value. When the wind generator system becomes stable, the switch would close,

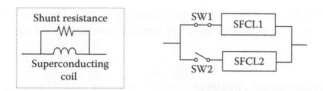

Figure 8.10 SFCL model.

and the circuit would be restored to its normal state. Figure 8.9 shows the control methodology of the braking resistor. According to the control method, if ΔV (difference in voltage) is positive, then the bypass switch is open; however, if ΔV is negative or zero, then the bypass switch is closed. Thus, a closed-loop control of the braking resistor is realized. It is important to note here that the bypass switch is based on the thyristor-controlled technology.

8.7   Superconducting Fault Current Limiter Method

A variety of SFCLs with various approaches to limiting current have been developed and tested. The FCL conceived of in this book consists of a detector, a controller, and a limiting resistance, all common hardware found in an FCL of any type.

An SFCL is employed as a superconducting-to-normal (S/N) transforming device, which uses a shunt resistance as shown in Figure 8.10. In the event of a fault, the superconductor undergoes a transition into its normal state (i.e., quenching). Because the superconducting coil needs a short time to recover normal operating condition after a quench, two SFCL devices connected in parallel are employed. Thus, the SFCL shown in Figure 8.1 actually consists of two SFCLs—namely, SFCL1 and SFCL2, as shown in Figure 8.10. Following a fault at first SFCL1 works, and when the circuit breaker is opened SW2 is closed just at the same time and hence current flows in the superconducting coil of SFCL2. In this book, 0.5 pu value of the limiting resistance is considered to obtain the best system performance.

8.8   Stabilization Methods Compared

The previous discussion revealed five methods of stabilization for fixed-speed wind generators that are recently being emphasized. However, although the individual technologies are well documented, a comparative study of these systems has not been reported so far. This book aims to fill in the gap and provides a comprehensive analysis of these stabilization methods for fixed-speed wind generator systems. The analysis is performed in terms of transient stability enhancement, controller complexity, and cost.

8.8.1   Performance Analysis

Effectiveness of each stabilization method is demonstrated through simulations performed using the electromagnetic transients program (EMTP) [37]. Simulations are carried out considering two cases: (1) a balanced (three-phase-to-ground, 3LG) fault occurs at point F1 near the synchronous generator at line 2 as shown in the system model of Figure 8.1; and (2) unsuccessful reclosure of circuit breakers occurs due to a permanent fault at point F1 near the synchronous generator at line 2. The time step and simulation time have been chosen as 50.0 μsec and 10.0 sec, respectively.

To clearly understand the effect of the stabilization methods, several performance indices, namely, *vlt (pu.sec)*, *spd (pu.sec)*, *pow (pu.sec)*, and *ang (deg.sec)*, as shown in Equations (8.2), (8.3), (8.4), and (8.5), respectively, are considered.

$$vlt\,(pu.\text{sec}) = \int_0^T \left| \Delta V \right| dt \qquad (8.2)$$

$$spd\,(pu.\text{sec}) = \int_0^T \left| \Delta \omega_r \right| dt \qquad (8.3)$$

$$pow\,(pu.\text{sec}) = \int_0^T \left| \Delta P \right| dt \qquad (8.4)$$

$$ang\,(deg.\text{sec}) = \int_0^T \left| \Delta \delta \right| dt \qquad (8.5)$$

In Equations (8.2) to (8.5), ΔV, $\Delta \omega_r$, ΔP, and Δd denote the terminal voltage deviation of the wind generator, the speed deviation of the wind generator, the real power deviation of the wind generator, and the load angle deviation of the synchronous generator, respectively, and T is the simulation time of 10.0 sec. The lower the values of the indices, the better the system's performance.

In this case, it is considered that the fault occurs at 0.1 sec, circuit breakers on the faulted lines are opened at 0.2 sec, and circuit breakers are closed again at 1.0 sec. It is assumed that the circuit breaker clears the line when the current through it crosses the zero level. Although actually the

Table 8.2 Values of Indices for Stabilization Methods
during Successful Reclosing

Index Parameters	Values of Indices					
	Pitch Method	BR Method	STATCOM Method	SMES Method	SFCL Method	Without Controller
vlt (pu.sec)	1.73	0.32	0.26	0.22	0.35	4.46
spd (pu.sec)	0.48	0.02	0.02	0.02	0.06	7.81
pow (pu.sec)	2.86	0.10	0.18	0.16	0.31	4.63
ang (deg.sec)	75.04	56.45	48.54	46.00	42.12	103.05

wind speed is randomly varying, during the short time span of the analysis of the transient stability the variation of wind speed can be considered negligible, and it is therefore assumed in this chapter that the wind speed is constant at 11.8 m/s.

Table 8.2 shows the values of the performance indices when the circuit breakers are successfully reclosed. It is seen that all methods are effective in transient stability enhancement; however, from the viewpoint of the index *pow (pu.sec)* the BR is the best, whereas with respect to the index *spd (pu.sec)* the BR, STATCOM, and SMES exhibit the same or better performance than the SFCL. From the perspective of *vlt (pu.sec)* the performance of the SMES is the best and the STATCOM is better than the BR and the SFCL. From the perspective of *ang (deg.sec)* the performance of the SFCL is the best, the SMES is better than the STATCOM and the BR, and the STATCOM is better than the BR. The pitch method exhibits the worst performance with respect to all indices.

Figure 8.11 shows the responses of the IG terminal voltage. It is seen that the IG terminal voltage returns back to its steady-state value due to the use of any of the devices of the SMES, STATCOM, BR, SFCL, and pitch controller. Figure 8.12 shows the responses of the IG rotor speed. It is seen that because of the use of any of the devices of the SMES, STATCOM, BR, SFCL, and pitch controller, IG becomes stable. Figure 8.13 shows the responses of the IG real power. In this case it is seeln that any of the devices of the SMES, STATCOM, BR, SFCL, and pitch controller can maintain the IG real power at the rated level. Figure 8.14 shows the responses of the SG load angle. It is clearly seen that the synchronous generator is transiently stable when any of the devices of the SMES, STATCOM, BR, SFCL, and pitch controller is used. This fact also indicates that the SMES, STATCOM, BR, SFCL, and pitch controller can make the entire power system stable when circuit breakers are successfully reclosed.

However, although SMES, STATCOM, BR, SFCL, and pitch controller can each make the wind generator stable, it is evident from the simulation results that the performance of the SMES is the best. The SFCL, STATCOM,

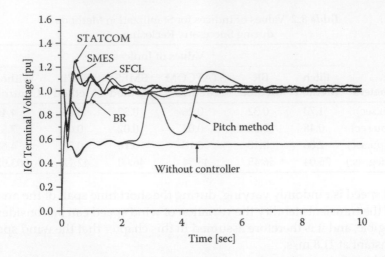

Figure 8.11 Responses of IG terminal voltage.

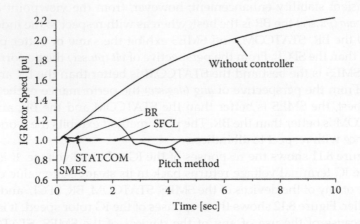

Figure 8.12 Responses of IG rotor speed.

and BR provide almost the same performance. The response of the pitch controller is much slower than that of the SMES, STATCOM, SFCL, and BR.

During the simulation, it is considered that the reclosing of circuit breakers is unsuccessful due to a permanent fault. Therefore, the circuit breakers are reopened 0.1 sec after they are reclosed. The values of the performance indices are shown in Table 8.3. From the indices it is seen that the SMES, STATCOM, and BR can stabilize both the wind generator and synchronous generator. In other words, these three methods can stabilize the overall system well. The pitch method can stabilize the wind generator but not the synchronous generator as evident from the index *ang (deg.*

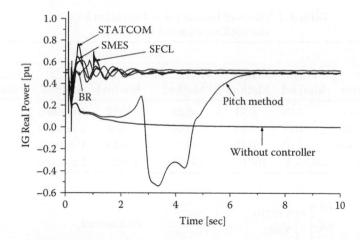

Figure 8.13 Responses of IG real power.

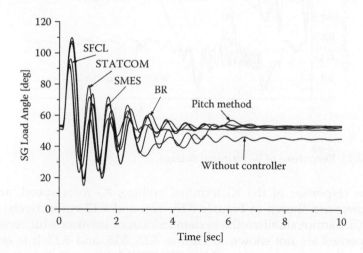

Figure 8.14 Responses of SG load angle.

sec); that is, the pitch method cannot stabilize the overall system in case of unsuccessful reclosing of circuit breakers. Also, from the perspective of the index *ang (deg.sec)* the performance of the SMES is the best, whereas from the viewpoints of *vlt (pu.sec)* and *pow (pu.sec)* the performance of the BR is the best and the SMES is better than STATCOM. From the viewpoint of *spd (pu.sec)* the performance of the SMES, STATCOM, and BR is the same. Again, with respect to all indices, the performance of the pitch method is worse than that of the SMES, STATCOM, and BR. It is seen that the SFCL cannot stabilize the system in case of unsuccessful reclosing.

Table 8.3 Values of Indices for Stabilization Methods
during Unsuccessful Reclosing

Index Parameters	Values of Indices					
	Pitch Method	BR Method	STATCOM Method	SMES Method	SFCL Method	Without Controller
vlt (pu.sec)	3.09	0.41	0.55	0.51	Unstable	5.22
spd (pu.sec)	0.67	0.04	0.04	0.04	Unstable	8.04
pow (pu.sec)	2.61	0.20	0.28	0.23	Unstable	4.70
ang (deg.sec)	24045.36	164.05	134.11	115.80	Unstable	28779.39

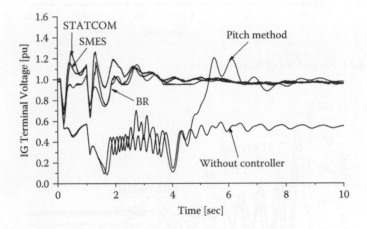

Figure 8.15 Responses of IG terminal voltage.

The responses of the IG terminal voltage, IG rotor speed, and IG real power are shown in Figures 8.15, 8.16, and 8.17, respectively. Since the SFCL cannot stabilize the system in case of unsuccessful reclosing, its responses are not shown in Figures 8.15, 8.16, and 8.17. It is evident that any of the devices of the SMES, STATCOM, BR, and pitch controller can stabilize the wind generator system in case of unsuccessful reclosing of circuit breakers. However, the performance of the SMES is the best, whereas the pitch controller exhibits the worst performance. Also, the response of the pitch controller is much slower than that of the SMES, STATCOM, and BR. The performance of the STATCOM and BR is almost the same. Figure 8.18 shows the responses of the SG load angle. It is seen that the pitch controller cannot stabilize the synchronous generator, but the SMES, STATCOM, and BR can make the synchronous generator transiently stable. In other words, the pitch method cannot stabilize the overall system, whereas the other three methods can make the entire power system stable.

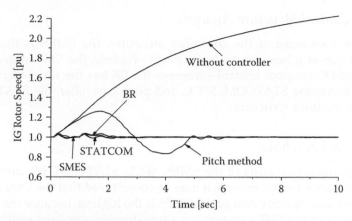

Figure 8.16 Responses of IG rotor speed.

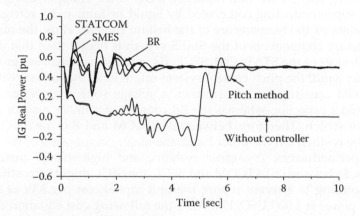

Figure 8.17 Responses of IG real power.

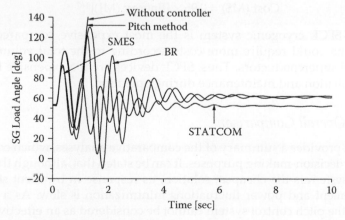

Figure 8.18 Responses of SG load angle.

8.8.2   Control Structure Analysis

From the viewpoint of the controller structure, the SMES is the most complex one as it has two control aspects—namely, the VSC control and the DC-to-DC chopper control—whereas the BR has the simplest control structure. Among STATCOM, SFCL, and pitch controller, the STATCOM has more complex structure.

8.8.3   Cost Analysis

Although the actual costs of the SMES, SFCL, STATCOM, BR, and pitch control systems are not known, it may be conjectured that the total installation and maintenance cost of the SMES is the highest, because the major components of the SMES system are a transformer, a voltage source converter using IGBT, a DC link capacitor, a DC-to-DC chopper using IGBT, a large superconducting coil cooled by liquid helium, and a refrigerator that maintains the temperature of the helium coolant. Thus, the number of necessary components of the SMES system is bigger than that of any of the devices of the STATCOM, SFCL, BR, and pitch control systems. On the other hand, the pitch control system might have the lowest cost. The STATCOM consists of a transformer, a voltage source converter using IGBT, and a capacitor, whereas the BR consists of a linear resistor and a thyristor switch. Therefore, between STATCOM and BR, the STATCOM might be costlier. On the other hand, the major components of the SFCL are superconductors, cryogenic systems, and high-voltage insulation devices. So between STATCOM and SFCL, the SFCL might be costlier.

According to a recent report, the unit capital cost per KW of SMES output power is 2,000 USD/kW. Also, the following cost equation can be applied to the SMES made of solenoid type of magnets [41]:

$$\text{Cost (M\$)} = 0.95 \times [\text{Energy (MJ)}]^{0.67} \qquad (8.6)$$

The SFCL cryogenic system is the most expensive compared with others and could require more costs according to the total volume and length of superconductors. Thus, SFCL devices are too expensive for initial installation and maintenance during operation.

8.8.4   Overall Comparison

Table 8.4 provides a summary of the comparative analyses, which could be used for decision-making purposes. It can be stated that, although the pitch control system is the cheapest solution, its response in transient stability enhancement and power fluctuations minimization is slow. As a consequence, the pitch control system cannot be considered as an effective solution. The braking resistor can be considered a very simple and cost-effective

Table 8.4 Overall Comparison of Stabilization Methods

Criteria	Pitch Control	BR	STATCOM	SMES	SFCL
			Stabilization Methods		
Ability to control active and reactive powers	Can control only active power	Can consume only active power	Can control only reactive power	Can control both active and reactive powers	Can control only active power
Transient stability enhancement during successful reclosing	Can stabilize the overall system, but its response is slower compared to BR, STATCOM, and SMES	Can stabilize the overall system and effective	Can stabilize the overall system and effective	Can stabilize the overall system and the most effective	Can stabilize the overall system and effective
Transient stability enhancement during unsuccessful reclosing	Can stabilize the wind generator but cannot stabilize the synchronous generator, that is, cannot stabilize the overall system	Can stabilize the overall system and effective	Can stabilize the overall system and effective	Can stabilize the overall system and the most effective	Cannot stabilize the wind generator system
Minimization of power and voltage fluctuations	Able to minimize only power fluctuations	Not able to minimize power and voltage fluctuations	Able to minimize only voltage fluctuations	Able to minimize both power and voltage fluctuations	Not able to minimize power and voltage fluctuatios
Controller complexity	More complex than BR	Simplest	More complex than pitch system	Most complex	More complex than BR
Manufacturing cost	Cheapest	Costlier than pitch control	Costlier than BR	Most expensive	Expensive

solution from the viewpoints of the transient stability enhancement of wind generator systems for both successfully and unsuccessfully reclosing circuit breakers. From the perspective of transient stability enhancement as well as voltage fluctuations minimization, STATCOM provides a cost-effective solution. SFCL is a good solution to stabilize the system to successfully reclose circuit breakers; however, it is costly. The SMES is the most expensive device; however, from the viewpoints of transient stability enhancement and minimization of both power and voltage fluctuations, it is the most effective solution. It is worth noting that, due to its salient properties such as very fast response, high efficiency, and capability of control of real power and reactive power, the SMES system is gradually gaining interest in the field of power systems. It is hoped that its potential advantages and environmental benefits will make SMES units a viable alternative for energy storage and management devices in the future. And although at present the cost of an SMES unit appears somewhat high, continued research and development are likely to bring the price down and make the technology appear even more attractive [40].

It is important to note here that the main challenge to implement the SMES and SFCL in a real system is their high cost. If the total implementation cost could be reduced, then both technologies could be made attractive. Thus, much more research is needed on how the total installation and operation cost of the SMES and SFCL could be reduced. Here one question might arise regarding why the costly SMES and SFCL devices are considered for wind generator stabilization, whereas the less costly methods like STATCOM, BR, and pitch control methods are available. The STATCOM method can control only reactive powers, and the BR and pitch control methods can control only active power. Thus none of these three methods (STATCOM, BR, and pitch control) can control simultaneously the real and reactive powers. The SMES has the ability to control both the real and reactive powers simultaneously and quickly. Again, SFCL can provide a very good solution to reduce the higher level of short-circuit current during a fault, which is increased by the rapid development of the WTGS. Therefore, in spite of their higher costs, the SMES and SFCL have been considered for wind generator stabilization.

One point to note here is that in this book the controller parameters have been determined by the trial-and-error approach, and the parameters are tuned very carefully so that the best system performance can be obtained. Although the results corresponding to only the severest fault case (3LG fault) are shown in this chapter, the designed parameters are tested in case of other faults also, like double-line-to-ground (2LG) fault, line-to-line (2LS) fault, and single-line-to-ground (1LG) fault in the system, and it is found that the system performance is good and effective. Thus, it can be emphasized that the designed controller parameters are robust and stable.

8.9 Chapter Summary

This chapter provides a thorough description on the methods of transient stability enhancement of wind generator systems. A comparative analysis of SMES, SFCL, STATCOM, braking resistor, and pitch control methods on the basis of transient stability enhancement, controller complexity, and manufacturing cost of fixed-speed wind energy systems is done. An important feature of this chapter is that the transient stability analysis is conducted considering unsuccessful reclosing of circuit breakers. It can be concluded that:

1. The pitch control system is the cheapest solution for wind generator stabilization in case of successful reclosing of circuit breakers and power fluctuations minimization, but its response is slow. As a consequence, the pitch control system cannot be considered as an effective stabilization means for wind generator systems.
2. The braking resistor method can be considered a very simple and cost-effective solution from the viewpoints of the transient stability enhancement of wind generation systems in both successfully and unsuccessfully reclosing circuit breakers.
3. From the perspective of transient stability enhancement in both successfully and unsuccessfully reclosing circuit breakers and voltage fluctuations minimization, STATCOM provides a cost-effective solution.
4. SFCL is an effective solution for transient stability enhancement in successfully reclosing circuit breakers, but it cannot stabilize the system during unsuccessful reclosing. Also, SFCL might be a costly device.
5. SMES is the most expensive device. However, from the viewpoints of transient stability enhancement in both successfully and unsuccessfully reclosing circuit breakers and minimizing both power and voltage fluctuations, the SMES system is the most effective solution.

This chapter helps readers understand the relative effectiveness of the stabilization methods and provides a guideline for selecting a suitable technique for the stabilization of wind energy systems.

References

1. M. H. Ali and B. Wu, "Comparison of stabilization methods for fixed-speed wind generator systems," *IEEE Transactions on Power Delivery*, vol. 25, no. 1, pp. 323–331, January 2010.
2. A. Sumper, O. G. Bellmunt, A. S. Andreu, R. V. Robles, and J. R. Duran, "Response of fixed speed wind turbines to system frequency disturbances," *IEEE Trans. Power Systems*, vol. 24, no. 1, pp. 181–192, February 2009.

3. G. S. Stavrakakis and G. N. Kariniotakis, "A general simulation algorithm for the accurate assessment of isolated diesel-wind turbines systems interaction. Part I: A general multimachine power system model," *IEEE Trans. Energy Conversion*, vol. 10, no. 3, pp. 577–583, September 1995.

4. G. S. Stavrakakis and G. N. Kariniotakis, "A general simulation algorithm for the accurate assessment of isolated diesel-wind turbines systems interaction. Part II: Implementation of the algorithm and case-studies with induction generators," *IEEE Trans. Energy Conversion*, vol. 10, no. 3, pp. 584–590, September 1995.

5. J. Tamura, T. Yamazaki, M. Ueno, Y. Matsumura, and S. Kimoto, "Transient stability simulation of power system including wind generator by PSCAD/EMTDC," 2001 *IEEE Porto Power Tech. Proceedings*, vol. 4, EMT-108, 2001.

6. E. S. Abdin and W. Xu, "Control design and dynamic performance analysis of a wind turbine-induction generator unit," *IEEE Trans. Energy Conversion*, vol. 15, no. 1, pp. 91–96, March 2000.

7. Z. S. Saoud, M. L. Lisboa, J. B. Ekanayake, N. Jenkins, and G. Strbac, "Application of STATCOMs to wind farms," *IEE Proc.-Gener. Transm. Distrib.*, vol. 145, no. 5, pp. 511–516, September 1998.

8. M. Aten, J. Martinez, and P. J. Cartwright, "Fault recovery of a wind farm with fixed speed induction generators using a STATCOM," *Wind Engineering*, vol. 29, no. 4, pp. 365–375, 2005.

9. H. Gaztanaga, I. E. Otadui, D. Ocnasu, and S. Bacha, "Real-time analysis of the transient response improvement of fixed-speed wind farms by using a reduced-scale STATCOM prototype," *IEEE Trans. Power Systems*, vol. 22, no. 2, pp. 658–666, May 2007.

10. M. H. Ali, T. Murata, and J. Tamura, "Effect of coordination of optimal reclosing and fuzzy controlled braking resistor on transient stability during unsuccessful reclosing," *IEEE Trans. Power Systems*, vol. 21, no. 3, pp.1321–1330, August 2006.

11. M. H. Ali, T. Murata, and J. Tamura, "The effect of temperature rise of the fuzzy logic-controlled braking resistors on transient stability," *IEEE Trans. Power Systems*, vol. 19, no. 2, pp. 1085–1095, May 2004.

12. A. Causebrook, D. J. Atkinson, and A. G. Jack, "Fault ride-through of large wind farms using series dynamic braking resistors (March 2007)," *IEEE Trans. Power Systems*, vol. 22, no. 3, pp. 966–975, August 2007.

13. W. Freitas, A. Morelato, and W. Xu, "Improvement of induction generator stability using braking resistors," *IEEE Trans. Power Systems*, vol. 19, no. 2, pp. 1247–1249, May 2004.

14. X. Wu, A. Arulampalam, C. Zhan, and N. Jenkins, "Application of a static reactive power compensator (STATCOM) and a dynamic braking resistor (DBR) for the stability enhancement of a large wind farm," *Wind Engineering*, vol. 27, no. 2, pp. 93–106, 2003.

15. A. Arulampalam, M. Barnes, N. Jenkins, and J. B. Ekanayake, "Power Quality and stability improvement of a wind farm using STATCOM supported with hybrid battery energy storage," *IEE Proc.-Gener. Transm. Distrib.*, vol. 153, no. 6, pp. 701–110, November 2006.

16. K. Rajambal, B. Umamaheswari, and C. Chellamuthu, "Electrical braking of large wind turbines," *Renewable Energy*, vol. 30, Issue 15, pp. 2235–2245, December 2005.

17. P. M. Anderson and A. Bose, "Stability simulation of wind turbine systems," *IEEE Trans. Power Apparatus and Systems*, vol. PAS-102, no. 12, pp. 3791–3795, December 1983.
18. J. Tamura, T. Yamazaki, R. Takahashi, S. Yonaga, Y. Matsumura, and H. Kubo, "Analysis of transient stability of wind generators," *Proceedings, of the International Conference on Electrical Machines (ICEM)* 2002, no. 148, 2002.
19. J. G. Slootweg, S. W. D. de Haan, H. Polinder, and W. L. Kling, "General model for representing variable speed wind turbines in power system dynamics simulations," *IEEE Trans. Power Systems*, vol. 18, no. 1, pp. 144–151, February 2003.
20. H. J. Boenig and J. F. Hauer, "Commissioning tests of the Bonneville Power Administration 30 MJ superconducting magnetic energy storage unit," *IEEE Trans. Power Apparatus and Systems*, vol. PAS-104, no. 2, pp. 302–309, February 1985.
21. S. C. Tripathy, M. Kalantar, and R. Balasubramanian, "Dynamics and stability of wind and diesel turbine generators with superconducting magnetic energy storage on an isolated power system," *IEEE Trans. Energy Conversion*, vol. 6, no. 4, pp. 579–585, December 1991.
22. S. Nomura, Y. Ohata, T. Hagita, H. Tsutsui, S. Tsuji-Iio, and R. Shimada, "Wind farms linked by SMES systems," *IEEE Trans. Applied Superconductivity*, vol. 15, no. 2, pp. 1951–1954, June 2005.
23. T. Kinjo, T. Senjyu, N. Urasaki, and H. Fujita, "Terminal-voltage and output-power regulation of wind-turbine generator by series and parallel compensation using SMES," *IEE Proc.-Gener. Transm. Distrib.*, vol. 153, no. 3, pp. 276–282, May 2006.
24. M. H. Ali, T. Murata, and J. Tamura, "Minimization of fluctuations of line power and terminal voltage of wind generator by fuzzy logic-controlled SMES," *International Review of Electrical Engineering (IREE)*, vol. 1, no. 4, pp. 559–566, October 2006.
25. M. H. Ali, T. Murata, and J. Tamura, "Wind generator stabilization by PWM voltage source converter and chopper controlled SMES," *CD record of ICEM (International Conference on Electrical Machines)* 2006, September 2006.
26. B. W. Lee, J. Sim, K. B. Park, and I. S. Oh, "Practical application issues of superconducting fault current limiters for electric power systems," *IEEE Trans. Applied Superconductivity*, vol. 18, no. 2, pp. 620–623, June 2008.
27. L. Ye, L. Lin, and K.-P. Juengst, "Application studies of superconducting fault current limiters in electric power systems," *IEEE Trans. Applied Superconductivity*, vol. 12, no. 1, pp. 900–903, March 2002.
28. B. C. Sung, D. K. Park, J.-W. Park, and T. K. Ko, "Study on optimal location of a resistive SFCL applied to an electric power grid," *IEEE Trans. Applied Superconductivity*, vol. 19, no. 3, pp. 2048–2052, June 2009.
29. B. C. Sung, D. K. Park, J.-W. Park, and T. K. Ko, "Study on a series resistive SFCL to improve power system transient stability: Modeling, simulation, and experimental verification," *IEEE Trans. Industrial Electronics*, vol. 56, no. 7, pp. 2412–2419, July 2009.
30. M. Tsuda, Y. Mitani, K. Tsuji, and K. Kakihana, "Application of resistor based superconducting fault current limiter to enhancement of power system transient stability," *IEEE Trans. Applied Superconductivity*, vol. 11, no. 1, pp. 2122–2125, March 2001.

31. W.-J. Park, B. C. Sung, and J.-W. Park, "The effect of SFCL on electric power grid with wind-turbine generation system," *IEEE Trans. Applied Superconductivity*, vol. 20, no. 3, pp. 1177–1181, June 2010.

32. L. Ye and L. Z. Lin, "Study of superconducting fault current limiters for system integration of wind farms," *IEEE Trans. Applied Superconductivity*, vol. 20, no. 3, pp. 1233–1237, June 2010.

33. J. Tamura, Y. Shima, R. Takahashi, T. Murata, Y. Tomaki, S. Tominaga, et al., "Transient stability analysis of wind generator during short circuit faults," *Keynote Lecture, Proc. Third International Conference on Systems, Signals & Devices (SSD'05)*, Sousse, Tunisia, March 21–24, 2005.

34. J. Tamura, S. Yonaga, Y. Matsumura, and H. Kubo, "A consideration on the voltage stability of wind generators," *Trans. IEE of Japan*, vol. 122-B, no. 10, pp. 1129–1130, October 2002.

35. S. Heier, *Grid integration of wind energy conversion system*, John Wiley & Sons, 1998.

36. IEEE Task Force on Benchmark Models for Digital Simulation of FACTS and Custom–Power Controllers, T&D Committee, "Detailed modeling of superconducting magnetic energy storage (SMES) system," *IEEE Trans. Power Delivery*, vol. 21, no. 2, pp. 699–710, April 2006.

37. *EMTP Theory Book*, Japan EMTP Committee, 1994.

38. T. Senjyu, R. Sakamoto, N. Urasaki, T. Funabashi, H. Fujita, and H. Sekine, "Output power leveling of wind turbine generator for all operating regions by pitch angle control," *IEEE Trans. Energy Conversion*, vol. 21, no. 2, pp. 467–475, June 2006.

39. Y.-S. Lee, "Decentralized suboptimal control of power systems with superconducting magnetic energy storage units," *International Journal of Power and Energy Systems*, vol. 21, no. 2, pp. 87–96, 2001.

40. P. D. Baumann, "Energy conservation and environmental benefits that may be realized from superconducting magnetic energy storage," *IEEE Trans. Energy Conversion*, vol. 7, no. 2, pp. 253–259, June 1992.

41. S. Nomura, T. Shintomi, S. Akita, T. Nitta, R. Shimada, and S. Meguro, "Technical and cost evaluation on SMES for electric power compensation," *IEEE Trans. Applied Superconductivity*, vol. 20, no. 3, pp. 1373–1378, June 2010.

42. H. Gaztanaga, I. E.-Otadui, D. Ocnasu, and S. Bacha, "Real-time analysis of the transient response improvement of fixed-speed wind farms by using a reduced-scale STATCOM prototype," *IEEE Trans. Power Systems*, vol. 22, no. 2, pp. 658–666, May 2007.

43. V. Vanitha, S. Shreyas, and V. Vasanth, "Fuzzy based grid voltage stabilization in a wind farm using static VAR compensator," Artcom, pp. 14–18, 2009 International Conference on Advances in Recent Technologies in Communication and Computing, 2009.

chapter 9

Fault Ride-Through Capability of Variable-Speed Wind Generator Systems

9.1 Introduction

Wind energy generation has been noted as the most rapidly growing renewable energy technology. The increasing penetration level of wind energy can have a significant impact on the grid, especially under abnormal grid voltage conditions. Thus, wind farms can no longer be considered as a simple energy source. Nowadays, they should provide an operational ability similar to that of conventional power plants. A demanding requirement for wind farms is the fault ride-through (FRT) capability. According to this demand, the wind turbine (WT) is required to survive during grid faults. The ability of a wind turbine to survive for a short duration of voltage dip without tripping is often referred to as the low-voltage ride-through (LVRT) capability of a turbine. An LVRT requirement for wind turbine systems is shown in Figure 9.1. On the other hand, power fluctuation from a turbine due to wind speed variations incurs a deviation of the system frequency from the rated value. As a result, it is necessary to mitigate this power fluctuation for high power quality. This chapter deals with the fault ride-through capability and mitigation of power fluctuations of variable-speed wind generator systems, especially for the doubly fed induction generator (DFIG), wound field synchronous generator, permanent magnet synchronous generator (PMSG), and synchronous reluctance generator systems [1–38]. Although the simulation results are not shown here, the technology and control algorithm of each of these variable-speed wind generators is presented.

9.2 Doubly Fed Induction Generator Systems

This section describes the ability of doubly fed induction generators to provide voltage stability support in weak transmission networks. Specifically, the response of wind turbines to voltage dips at the point of common coupling and its effects on system stability are analyzed. A control strategy for the operation of the grid and rotor side converters (RSCs) is developed to

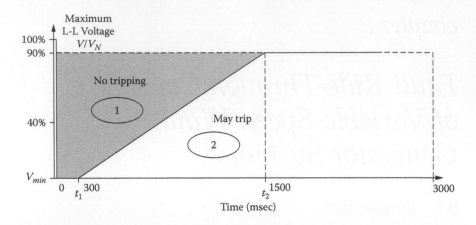

Figure 9.1 A grid code.

support the grid voltage by injecting reactive power during and after grid fault events. The performance of the strategy is analyzed for different voltage dips at the point of common coupling of a wind farm and compared with the case when the converters do not provide any voltage support [35].

DFIGs are the most common technology used in variable-speed wind turbines, having 45% of the medium to large wind turbines installed in Europe in 2005. In normal grid conditions, the use of power converters enables DFIGs to operate at optimal rotor speed and to maximize power generation by controlling the active and reactive power injected into the grid. When the voltage dips close to the wind farm, high currents will pass through the stator winding and will also flow through the rotor winding due to the magnetic coupling between stator and rotor. Such high currents could damage the converters; therefore, a protection system is required. The protection of the converter is usually achieved by short circuiting the generator rotor through a crowbar and thus blocking the rotor side converter. Once the rotor side converter is blocked, the DFIG operates like a typical induction generator, and therefore the control of active and reactive power through the rotor is inactive.

The aim of this section is to provide insight and understanding about the effective FRT capability of DFIGs in weak transmission networks and their effects on system stability. A control strategy allowing the grid and rotor side converters to support the grid voltage by injecting reactive power during and after grid faults is developed.

Figure 9.2 shows the arrangement of a DFIG. This concept uses a wound rotor induction generator whose stator windings are directly connected to the grid and whose rotor winding is connected to the network via a back-to-back insulated gate bipolar transistor (IGBT)-based converter. The rotor side converter regulates the active and reactive power

Figure 9.2 Doubly fed induction generator.

injected by the DFIG, and the grid side converter (GSC) controls the voltage at the direct current (DC) link.

The overall structure of the wind turbine model comprises the aerodynamic model, mechanical model, and the electrical model for the generator. The well-known actuator disk concept is taken into account by the aerodynamic model under the assumption of constant wind velocity. The drive train is approximated by a two-mass model considering one large mass representing the turbine rotor inertia and one small mass representing the generator rotor. The two masses are connected by a flexible low-speed shaft characterized by stiffness and damping. As usual in fundamental frequency simulations, the generator is represented by a third-order model whose equations are simplified by neglecting the stator transients.

A pitch angle control is also implemented to limit the generator speed during grid disturbances and in normal operation under high wind speeds. Finally, a protection system is included to block the rotor side converter when its safe operation is threatened. The protection system monitors the voltage at the point of common coupling (PCC), the magnitude of the rotor current, and the generator rotor speed. When at least one of these variables exceeds the range of its maximum and minimum values, the protection system blocks the rotor side converter by short circuiting the generator rotor through a crowbar.

9.2.1 Rotor Side Converter

The RSC controls independently the active and reactive power injected by the DFIG into the grid in a stator flux dq-reference frame. Figure 9.3 shows the control scheme of the RSC. The q-axis current component is used to control the active power using a maximum power tracking (MPT) strategy to calculate the active power reference. The reference value for the active power is compared with its actual value, and the error is sent to a proportional-integral (PI) controller that generates the reference value for the q-axis current. This signal is compared with its actual value, and the error is passed through a second PI controller determining the reference voltage for the q-axis component.

The d-axis is used to control the reactive power exchanged with the grid, which in normal operation is set to zero to operate with unity power factor. In case of disturbance, if the induced current in the rotor circuit

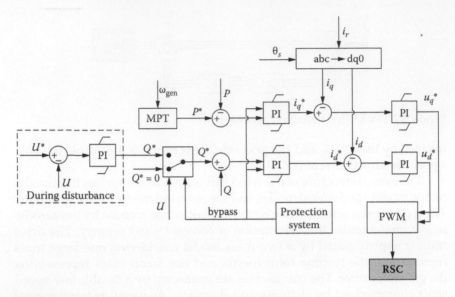

Figure 9.3 Control diagram of the rotor side converter.

is not high enough to trigger the overcurrent protection the RSC is set to inject reactive power into the grid to support the voltage restoration. In such cases, the actual voltage at the PCC is compared withits reference value, and the error is passed through a PI controller to generate the reference signal for the reactive power of the DFIG. Similar to the control strategy of the q-component, the error between the reactive power reference and its actual value is passed through a PI controller to determine the reference value for the d-axis current. This signal is compared with the d-axis current value, and the error is sent to a third PI controller that determines the reference voltage for the d-axis component. Finally, the dq-reference voltages are passed through the pulse width modulation (PWM) module, and the modulation indices for the control of the RSC are determined.

9.2.2   Grid Side Converter

The objective of the GSC is to maintain the voltage at the DC link between both power converters. In normal operation, the RSC already controls the unity power factor operation, and therefore the reference value for the exchanged reactive power between the GSC and the grid is set to zero. In case of disturbance, the GSC is set to inject reactive power into the grid whether the RSC is blocked or is kept in operation. Figure 9.4 shows the control diagram of the GSC.

As for the RSC, the control of the GSC is performed using the dq-reference frame, but instead of rotating with the stator flux the axis rotates with

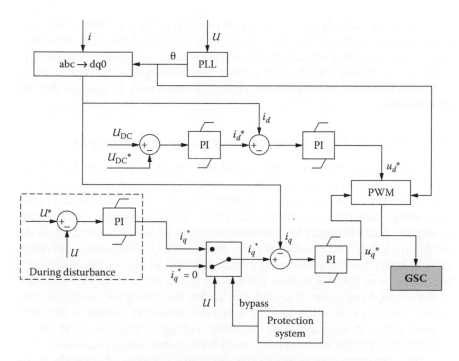

Figure 9.4 Control diagram of the grid side converter.

the grid voltage. The actual voltage at the DC link is compared with its reference value, and the error between both signals is passed through a PI controller that determines the reference signal for the d-axis current. This latter signal is subtracted from its current value, and the error is sent to another PI controller to obtain the reference voltage for the d-axis component.

As for the q-axis current, its reference value depends on whether the system operates in normal operation or during disturbance. In normal operation, the GSC is assumed reactive neutral by setting the reference value of the q-axis current to zero. In case of disturbance, the actual AC side voltage of the GSC is compared with its reference value, and the error is passed through a PI controller that generates the reference signal for the q-axis current. This reference signal is compared with its current value, and the error is sent to a second PI controller that establishes the reference voltage for the q-axis component. Finally, both reference voltages in a dq-reference frame are sent to the PWM module, which generates the modulation indices for the control of the GSC.

The injection of active and reactive power by the GSC is limited by its nominal capacity represented by the following equation in per-unit base:

$$|I_{conv}| = \sqrt{(I_q)^2 + (I_d)^2} \leq 1 \tag{9.1}$$

This work considers a strategy that prioritizes the injection of reactive power (q-axis current). The d-axis current is calculated based on Equation (9.1). During normal operation, the strategy does not present limitations with the control of the DC link voltage since the q-axis current is set to zero and therefore the converter capacity is used only to control the DC link voltage.

9.3   *Wound Field Synchronous Generator Systems*

This section describes a variable-speed WT based on a multipole, wound field synchronous generator. Due to the low generator speed, the rotor shaft is coupled directly to the generator. The generator is connected to the grid via an alternating current (AC)-to-DC-to-AC converter cascade and a step-up transformer. The converter consists of an uncontrolled diode rectifier, a DC-to-DC boost converter, and a PWM voltage source inverter (VSI). The layout of the electrical part is depicted in Figure 9.5. Major control systems of the WT system are shown in Figure 9.6. They include the turbine speed controller, the pitch controller, the generator excitation controller, and the GSC controls. The generator terminal voltage is rectified to feed the field winding. The excitation voltage is regulated via the Ef controller in Figure 9.6, which has sufficient overexcitation capability [36].

The converter connected to the generator consists of a diode rectifier, a boost DC-to-DC converter, and a PWM VSI. The boost converter comprises an inductor, an IGBT switch, a diode, and the output capacitor, and its purpose is to control the rectifier output current and therefore the generator current and torque. At the same time, it interfaces the rectified

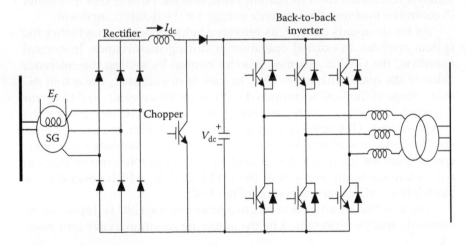

Figure 9.5 Electrical scheme of a variable speed wind turbine equipped with a direct-drive synchronous generator.

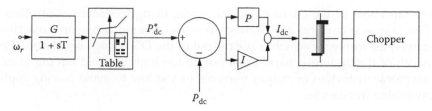

(a) Speed controller

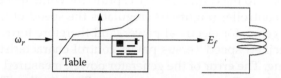

(b) Excitation controller

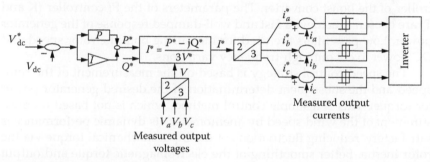

(c) Grid-side inverter controller

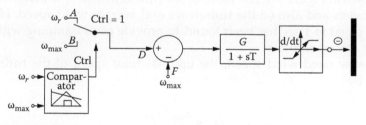

(d) Pitch controller

Figure 9.6 Controllers of the wind turbine.

generator voltage, which varies with speed, to the constant voltage at the input of the grid side inverter. The DC chopper is current controlled using a simple hysteretic controller. The grid side inverter is a standard three-phase two-level unit, consisting of six IGBTs and antiparallel diodes. It operates also in the current control mode, using hysteresis controllers. The reference currents are derived from the desired active and reactive power, the former of which is the output of the DC voltage regulator. The width of the hysteresis band must not be too high to avoid having great harmonic distortion of output currents or too low to avoid having high switching frequencies.

9.3.1   Speed Controller

The goal of speed control is to maximize energy capture for wind speeds below rated. The speed controller (Figure 9.6) regulates the speed of the rotor by controlling the generator electrical power (and therefore torque) according to the prespecified speed versus power control characteristic discussed in the following. The error of the generator power, measured at the DC side, determines the DC current set point via a PI controller. The DC current is then regulated at the reference value by the hysteresis controller of the boost converter. The parameters of the PI controller (K and T) are chosen to achieve a fast and well-damped response of the generator power. Low-pass filters are used for measuring both the rotor speed and DC power to cut out high-frequency variations.

The speed control strategy is based on the measurement of the rotor speed and the subsequent determination of the desired generator power (or torque). This is a simple control method, which is not based on measurement of the wind speed by anemometer. Its dynamic performance is satisfactory, reducing fluctuations of the input mechanical torque via the rotor inertia. Better smoothing of the electromagnetic torque and output power is achieved using a low-pass filter for the measured rotor speed, as shown in Figure 9.6. The value of the time constant, T, depends on WT parameters and also on the turbulence and average wind speed. Here, a value equal to 1 sec has been found to provide good damping with sufficient control accuracy.

Below rated wind speed, the optimal rotor speed of the turbine is given by

$$\omega_m = \frac{\lambda_{opt} * V_W}{R} \tag{9.2}$$

The optimal mechanical power for each specific wind speed is then calculated from Equation (2.5) for $C_p = C_{p,opt} = 0.44$ and shown in Figure 9.7 as a function of the corresponding optimal rotor speed. The range of rotor speed variation is approximately 13–43 rpm. The diagram in Figure 9.7 constitutes the control characteristic implemented in the speed controller.

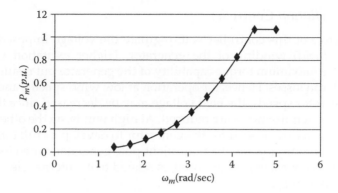

Figure 9.7 Speed control characteristic.

Power losses in the drive train, generator, and rectifier have been taken into account in this characteristic. The cut-in wind speed is considered equal to 4 m/s.

9.3.2 Pitch Controller

While speed control is effective below the rated wind speed to achieve maximum energy capture, at higher winds the goal of the WT control system is to maintain rated output power without rotor overspeed. This is achieved by the pitch controller, which is active only at high wind speeds. Then the generator power is regulated at its rated value to avoid overloading the generator and converters, whereas the blade pitch angle is increased to reduce C_p and the resulting mechanical torque. The objective is to prevent the rotor speed from increasing beyond its maximum value. For operation below rated wind speed, the pitch angle is approximately zero for maximum aerodynamic efficiency.

In Figure 9.6, a simple implementation of the pitch controller is shown as implemented in this work. When the rotor speed is lower than its maximum value (corresponding to the rated power on the control characteristic of Figure 9.7), the control switch is at position B, forcing a zero error ($\omega_{max} - \omega_{max} = 0$) and hence zero pitch angle output. Otherwise, the positive error drives the pitch mechanism to higher angles. A rate limiter, set at 5 deg/s, appears on the diagram. The first-order lag represents the time delay to the pitch mechanism. Finally, a proportional controller is employed since small overspeeds can be allowed and because the system is never actually in steady-state and hence the advantage of zero steady-state error of an integral controller is not applicable. The value of the gain G has been selected via repeated simulations for good dynamic performance.

9.3.3   Excitation Controller

The objective of this controller is to regulate the voltage imposed to the excitation (field) winding of the generator. Higher excitation voltages increase the maximum torque capability of the generator but result also in increased iron losses. Hence, for operation at low wind speeds, and therefore at low rotor speeds, the field voltage may be decreased since the generator torque requirements are reduced. At high winds, on the other hand, high excitations are needed for the generator to develop its full torque.

The maximum torque of a nonsalient pole generator, when its terminal voltage is not externally imposed, is related to its internal electromagnetic field (EMF) E_f:

$$T_{e,max} = \frac{3.Ef^2}{2.X_d.\omega_m} \Rightarrow Ef = \sqrt{\frac{2.T_{e,max}.X_d.\omega_m}{3}} \qquad (9.3)$$

For a specific rotor speed ω_m and corresponding torque (based on the speed control characteristic) the minimum required E_f is found. A security factor, here equal to 15%, is then applied to account for inaccuracies of Equation (9.3) and dynamic considerations. From the minimum internal EMF value, the corresponding field winding voltage, V_r, is then found, taking into account that their p.u. values are related by $E_f = v_r \omega_m \cdot$(p.u.). The field voltage control characteristic obtained is shown in Figure 9.8.

9.3.4   Grid Side Inverter Controller

The objective of this controller is to regulate the output power of the WT when it is connected to the grid. It includes an inner current control loop based on the hysteresis controller that regulates the output

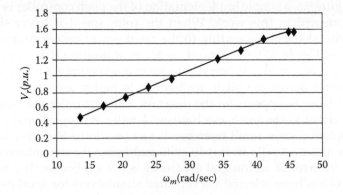

Figure 9.8 Speed versus field voltage control characteristic.

currents to their reference values, determined by the power control section, via static manipulations. The measured three-phase voltages at the output of the WT are transformed to the synchronous reference frame (rotating at the grid frequency). The reference output currents are then transformed in phase coordinates and used as inputs to the hysteresis current controllers. The active power set point P* is determined by the output of the DC voltage controller, implemented via a PI regulator, which accepts as input the error between the measured and the reference DC voltage.

9.4   Permanent Magnet Synchronous Generator Systems

With numerous advantages, PMSG systems represent an important trend in development of wind power applications. Extracting maximum power from wind and feeding the grid with high-quality electricity are two main objectives for wind energy conversion systems (WECSs). To realize these objectives, the AC-to-DC-to-AC converter is one of the best topologies for WECSs. Figure 9.9 shows a conventional configuration of an AC-to-DC-to-AC topology for a PMSG. This configuration includes a diode rectifier, a boost DC-to-DC converter, and a three-phase inverter. In this topology, the boost converter is controlled for maximum power point tracking (MPPT), and the inverter is controlled to deliver high-quality power to the grid [37].

Variable-speed wind turbines using a PMSG equipped with full-scale back-to-back converters are very promising and suitable for application in large wind farms. Due to their full-scale power converter, they can deliver a larger amount of reactive power to the grid than a DFIG wind turbine under abnormal grid conditions. To achieve LVRT capability for wind turbine systems, a braking chopper method was reported, which is a relatively cheap solution with a simple control. However, it is hard to

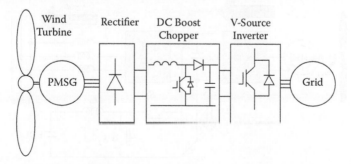

Figure 9.9 Conventional PMSG-based WECS with DC boost chopper.

smooth the fluctuated power from the turbine, and the generated power is dissipated in the braking resistor.

On the other hand, other methods using an energy storage system have been reported in the literature, but these methods are not suitable for LVRT since the energy and power capacity of the energy storage system (ESS) should be high enough to absorb the full differential power between the generator and the grid during the voltage sag. However, it can be applied to smooth power fluctuations. A study using an ESS for both LVRT and power smoothening has been reported, but the control algorithm for the ESS was not shown in detail and the power rating of the ESS was too high.

To reduce the minimum energy and power capacity of an ESS so that it can absorb the full differential power under grid voltage dips, the generator speed can be increased to store the kinetic energy in the system inertia. During this operation, the turbine output power extracted from the wind is not maximal since the generator speed is not the optimal value for the MPPT. In this chapter, a ride-through technique for PMSG wind turbine systems using an ESS is discussed. By storing the more inertial energy in the rotational body of the system by increasing the generator speed, the energy capacity of the ESS required can be reduced. In addition, an ESS integrated into PMSG wind turbine systems can also be used to improve the generator output power within the capacity of the ESS by charging or discharging the fluctuated component of the power under wind speed variations. The design procedure of the ESS is described in detail. The ESS consists of a DC-to-DC buck–boost converter and an electric double-layer capacitor (EDLC), which is connected at the DC link of the back-to-back converters as shown in Figure 9.10. A control strategy for the ESS composed of power and current controllers is suggested, resulting in an improvement in the overall performance for both ride-through and power smoothing.

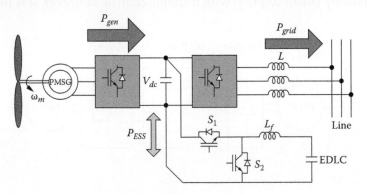

Figure 9.10 PMSG wind turbine system with ESS.

9.4.1   Control of Back-to-Back Converters

A line side converter (LSC) has a conventional cascaded control structure composed of an inner current control loop and an outer DC link voltage control. For vector control of a PMSG, a cascaded control scheme composed of an inner current control loop and an outer speed control loop is employed. To obtain maximum torque at a minimum current, the d-axis reference current component is set to zero and then the q-axis current is proportional to the active power, which is determined by the speed controller.

The control block diagram of a PMSG wind turbine is shown in Figure 9.11. The MPPT method is applied for turbine power control, which gives the speed reference of the PMSG under normal grid conditions. At a grid voltage sag, however, the MPPT control stops to reduce the turbine power extracted from wind. During this operation, the speed reference of the system is set higher than in the case of presag, which means that some portion of the turbine power can be stored in the system inertia.

The dynamic equations of the two-mass model of a PMSG wind turbine system without a gearbox, shown in Figure 9.12, are expressed as

$$T_t - T_k = j_t \frac{d\omega_t}{dt} \tag{9.4}$$

$$T_k - T_g = j_g \frac{d\omega_g}{dt} \tag{9.5}$$

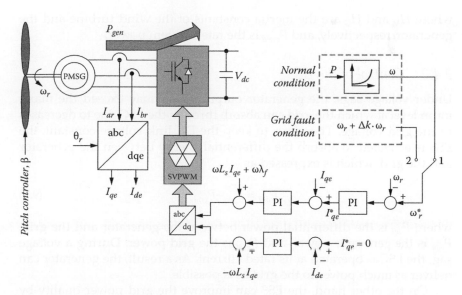

Figure 9.11 Control block diagram of PMSG.

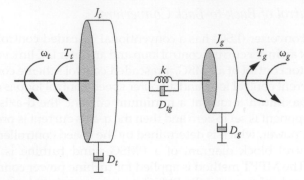

Figure 9.12 Two-mass model for drive train of wind turbine.

where T_t is the turbine torque, J_t is the inertia of the turbine, T_k and T_g are the torques in the flexible coupling and the generator, respectively, w_t and w_g are the mechanical speed of the turbine and the generator, respectively, and J_g is the inertia of the generator side.

Define the fault duration and the generator speed change as ΔT and $\Delta k(\%)$, respectively. From (9.4), (9.5), and the inertia constant of the system, the mechanical power, P_J, for speed variation is expressed as

$$P_J = 2P_{rated}(H_M + H_G)\frac{\Delta_K}{\Delta T} \tag{9.6}$$

where H_M and H_G are the inertia constants of the wind turbine and the generator, respectively, and P_{rated} is the rated system power.

9.4.2 Control of the ESS

Under voltage sags, the generator output power may exceed the maximum level at which the grid can absorb through the LSC due to decreases in the grid voltage. Therefore, to keep the DC link voltage constant, the ESS is activated to absorb the differential energy between the generator and the grid, which is expressed as

$$P_{diff} = P_{gen} - P_{grid} \tag{9.7}$$

where P_{diff} is the differential power between the generator and the grid, P_{gen} is the generator power, and P_{grid} is the grid power. During a voltage sag, the LSC is operated at its rated current. As a result, the generator can deliver as much power to the grid as is possible.

On the other hand, the ESS can improve the grid power quality by charging or discharging the fluctuated power component due to wind

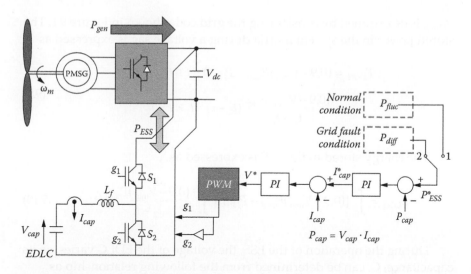

Figure 9.13 (See color insert.) Control block diagram for ESS.

speed variations. It is known that the power fluctuations in a particular frequency region (between 0.05 and 1 Hz) result in a system frequency deviation. Therefore, to eliminate the power fluctuations, the ESS needs to absorb or release the high-frequency component of the power P_{flux}, which is obtained by applying a second-order high-pass filter to the generator power so that

$$P_{flux} = \frac{S^2}{S^2 + 2\xi\omega_c S + \omega_c^2} P_{gen} \tag{9.8}$$

where x is the damping ratio, and w_c is the cutoff frequency ($w_c = 2\Pi f_c$). In this chapter, the cutoff frequency is chosen to be 0.1 Hz.

The role of the ESS is to absorb the power from the PMSG or to release it to the grid, as required. For this purpose, a power controller is used in the main control loop with an inner current control loop as shown in Figure 9.13. The power reference is given by (9.7) for ride-through and (9.8) for power smoothing.

9.4.3   Rating of the ESS

To meet the requirements of the grid code, the ESS should have the capability to absorb the full differential energy between the generator and the grid in a worst case scenario. By increasing the generator speed during a voltage sag, the ESS can be made small in size since some turbine output power is stored in the inertia of the system. The rated power for the LVRT,

P_{LVRT}, is determined by considering the grid code shown in Figure 9.1. The stored power in the system inertia during a voltage sag is expressed as

$$P_{LVRT} = (0.9 - V_{min})P_{rated} - P_J \quad 0 \le t \le t_1$$

$$= \frac{(0.9 - V_{min})P_{rated}}{t_2 - t_1}(t_2 - t) \quad t_1 < t \le t_2 \tag{9.9}$$

The energy stored in the ESS is expressed as

$$E_{LVRT} = \int_0^{t_1} [(0.9 - V_{min})P_{rated} - P_J]dt + \int_{t_1}^{t_2} \frac{(0.9 - V_{min})P_{rated}}{t_2 - t_1}(t_2 - t) \tag{9.10}$$

During the operation of the ESS, the voltage of the EDLC varies, so its capacitance, C, can be determined from the following relationship as

$$C = \frac{2.E_{LVRT}}{\Delta V_{cap}.V_{cap}^{rated}} \tag{9.11}$$

where V_{cap}^{rated} and ΔV_{cap} are the rated voltage and the voltage variation of the EDLC, respectively.

It is noted that this scheme can improve the grid power fluctuations. The main cause of power oscillations in wind turbines is the wind speed variation, which is expressed as

$$V(t) = V_{w0} + \sum \Delta V_{wi} \sin(\omega_i t) \tag{9.12}$$

where V_{w0} is the mean wind speed, ΔV_{wi} is the harmonic amplitude, and w_i is angular frequency ($f = 0.1 \sim 10$ Hz). In practice, wind speed variations are randomly varied and depend on the regional environment. It is assumed that the wind fluctuation is 30% of the mean value. Therefore, the power capacity of the ESS in this case is suitable for short-term energy storage as is shown clearly in the following design example.

9.4.4 *Design Example for the ESS*

The parameters of the ESS are designed for a 2 MW PMSG wind turbine system, where the total inertia constant of the wind turbine and the generator is given as 6 sec. According to the grid code, for a $\Delta T = 300$ ms in the LVRT curve, the system rotational speed is controlled to within $\Delta k =$

1.0% during this duration. Thus, the mechanical power stored in system inertia is, by (9.6),

$$P_J = 2.2[MW].6.\frac{0.01}{0.3} = 0.8[MW]$$

Then, the rated power of the ESS is calculated from the grid code and the P_J previously given as

$$P_{LVRT} = (0.9-0).2[MW]-0.8 = 1.0[MW]$$

Then, the rated energy capacity of the EDLC is given by

$$E_{LVRT} = \int_0^{0.3} 1.0 dt + \int_{0.3}^{1.5} \frac{1.8}{1.5-0.3}(1.5-t)dt = 1.38[MJ]$$

While the EDLC is charged or discharged, the voltage of the EDLC will vary when the voltage variation is set to 20% of the rated value. In this chapter, the rated voltage of the EDLC is set to 400 V. Therefore, the capacitance of the EDLC is determined as

$$C = \frac{2.1.38[MJ]}{400.80} = 86.25[F]$$

For power fluctuation at 0.3 p.u., with this EDLC capacity, the ESS can operate during the period T_{fluc} as

$$T_{fluc} = \frac{1.38[MJ]}{0.6[MW]} = 2.3[s]$$

9.5 Switched Reluctance Generator System

This section discusses the switched reluctance machine (SRM) application in the wind generator system. In the last decades the SRM has become an important alternative in various applications within both the industrial and domestic markets, namely, as a motor showing good mechanical reliability, high torque-volume ratio and high efficiency, plus low cost.

Although SRMs are less evangelized as a generator, there have been several studies regarding their application in the aeronautical industry and in integrated applications in wind-based energy generators. Although easy to build, the SRM in the past was a source of complaints concerning to its dynamic performance and the peculiar characteristics of its command and control. At the time, these arguments were sufficiently convincing to stop a sustained development and research of this kind of machine. The development of power electronics, and especially the advancements in the field of semiconductors, brought improvements in the command and control technology of this kind of machine, thus spearheading a diversified application of SRMs.

The principles of operation of this machine are simple, well known, and based on reluctance torque. The machine has a stator of wound-up salient poles that after energizing synchronized with the position of the rotor develops a torque that tends to align the poles in a way that diminishes the reluctance in the magnetic circuit. Currently the synchronous and induction machines dominate the market of wind energy applications, although the SRM has been the subject of current investigation and it shows to be a valid alternative for this field. Compared with the classical solutions of machines integrated in wind applications, a switched reluctance generator (SRG) shows a simplified construction associated with the inexistence of permanent magnets or conductors in the rotor, which results in lower manufacturing costs; in addition, both the machine and the power converter are robust. The low inertia of the rotor allows the machine to respond to rapid variations in the load. Associated with these characteristics, these machines have a control system that allows rapid changes in the control strategy such that the performance of the machine is optimized.

The structure of the SRM is not as stiff as the synchronous machines, and due to its flexible control system it is capable of absorbing transient conditions, thus supplying more resilience to the mechanical system. The machine has an inherent fault tolerance, especially when under an open-coil fault (in the windings) and in the power converter (external faults). Under normal operation, each phase of SRG is electrically and magnetically independent from others. The SRM is generally felt to be louder than conventional machines. However, an adequate mechanical design can do a lot to improve these figures, and new control techniques—current control strategy with a torque reference—permit further improvements.

9.5.1 SRG Operation

In electrical drives with variable reluctance, as shown in Figure 9.14, the torque is a function of the angular position of the rotor due to the double salient poles. The operation of this machine as a generator is obtained by energizing the windings of the stator when the salient poles of the rotor are away from their aligned position due to the rotating motion of the prime mover. A commercially available switched reluctance machine used in this study was a 2,4 kW four-phase 8/6 machine [38].

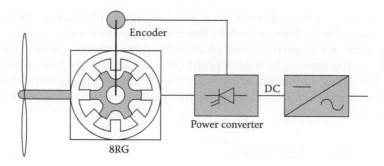

Figure 9.14 Switched reluctance generator in a wind turbine.

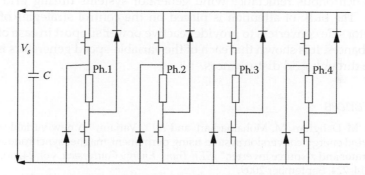

Figure 9.15 Circuit diagram of the four-phase converter for SRG.

The SRM, although being simple from the construction point of view, is characterized by a peculiar mode of controlling its phase currents. For that matter a power electronic converter is used that functions in a way that the phase currents of the machine are imposed for certain positions of the rotor. This study used a standard topology of the converter usually applied in SRM drives, given that it provides a greater flexibility regarding its control and better fault tolerance. Another reason of its reliability during fault conditions is the electrical independence among phases. The control system of this converter must regulate the magnitude and even the wave shapes of the phase currents to fulfill the requirements of torque and output power available and to ensure safe operation of the generator. This implies that the electronic switches associated with the controller are fully controlled devices. These devices work to invert the voltage applied to the phases in certain angular positions of the rotor and also assist in the commutations of phases. The topology shown in Figure 9.15 uses power transistors (IGBT or MOSFET) that work as electronic switches. The capacitor shown in the aforementioned topology prevents fluctuations in the voltage Vs.

The SRG is a valid alternative in wind energy applications. Therefore, it is reasonable to foresee that in the medium power wind systems the SRGs allow good performance in extracting the energy carried by the wind. On the downside we can point out the fact that the SRG is noisier than the other conventional systems. Nevertheless the current control based on torque reference attenuates this problem, especially via a reduction of the torque ripple.

9.6   Chapter Summary

This chapter discusses the fault ride-through capability of DFIG, wound field synchronous generator, permanent magnet synchronous generator, and synchronous reluctance wind generator systems during grid fault events. The bulk of attention is placed on the control strategies of grid and rotor side converters to provide reactive power support in case of grid disturbances. It is shown that each of the variable-speed generators is able to ride through grid disturbances.

References

1. S. M. Dehghan, M. Mohamadian, and A. Y. Varjani, "A new variable-speed wind energy conversion system using permanent-magnet synchronous generator and z-source inverter," *IEEE Trans. Energy Conversion*, vol. 24, no. 3, pp. 714–724, September 2009.
2. M. B. C. Salles, K. Hameyer, J. R. Cardoso, and W. Freitas, "Dynamic analysis of wind turbines considering new grid requirements," *IEEE International Conference on Electrical Machines*, Portugal, September 2008.
3. A. D. Hansen and G. Michalke, "Fault ride-through capability of DFIG wind turbines," *Renewable Energy*, vol. 32, no. 9, pp. 1594–1610, July 2007.
4. A.D. Hansen, P. Sørensen, F. Blaabjerg, and J. Bech, "Dynamic modeling of wind farm grid interaction," *Wind Engineering*, vol. 26, no. 4, pp. 191–208, 2002.
5. P. Kundur, *Power system stability and control*, McGraw-Hill, 1994.
6. Y. Lei, A. Mullane, G. Lightbody, and R. Yacamini, "Modeling of the wind turbine with a doubly fed induction generator for grid integration studies," *IEEE Trans. Energy Conversion*, vol. 21, pp. 257–264, March 2006.
7. N. Rahmat, T. Thiringer, and D. Karlsson, "Voltage and transient stability support by wind farms complying with the E.ON Netz grid code," *IEEE Trans. Power Systems*, vol. 22, no. 4, pp. 1647–56, November 2007.
8. P. Sorensen, N. A. Cutuluis, A. V. Rodriguez, L. E. Jensen, J. Hjerrild, M. H. Donovan, et al., "Power fluctuations from large wind farms," *IEEE Trans. Power Systems*, vol. 22, no. 3, pp. 958–1065, August 2007.
9. J. F. Conroy and R. Watson, "Low-voltage ride-through of a full converter wind turbine with permanent magnet generator," *IET Renew. Power. Gener.*, vol. 1, no. 3, pp. 182–189, September 2007.
10. C. Abbey and G. Joos, "Supercapacitor energy storage for wind energy applications," *IEEE Trans. Ind. App.*, vol. 43, no. 3, pp. 769–776, May–June 2007.

11. R. Data and V. T. Ranganathan, "A method of tracking the peak power points for a variable speed wind energy conversion system," *IEEE Trans. Energy Conversion*, vol. 18, no. 1, pp. 163–168, March 2003.

12. H- S. Song and K. Nam, "Dual current control scheme for PWM converter under unbalanced input voltage conditions," *IEEE Trans. Ind. App.*, vol. 46, no. 5, pp. 953–959, October 1999.

13. C. Luo, H. Banakar, B. Shen, and B. T. Ooi, "Strategy to smooth wind power fluctuation of wind turbine generator," *IEEE Trans. Energy Conversion*, vol. 22, no. 2, pp. 341–349, June 2007.

14. J. Morren, J. Pierik, and S. W. H. de Haan, "Inertial response of variable speed wind turbines," *Electric Power Systems Research*, vol. 76, no. 11, pp. 980–987, July 2006.

15. D. Xiang, L. Ran, P. J. Tavner, and S. Yang, "Control of a doubly fed induction generator in a wind turbine during grid fault ride-through," *IEEE Trans. Energy Conversion*, vol. 21, no. 3, pp. 652–662, September 2006.

16. A. M. Knight and G. E. Peters, "Simple wind energy controller for an expanded operating range," *IEEE Trans. Energy Conversion*, vol. 20, no. 2, pp. 459–466, June 2005.

17. J. G. Slootweg, S. W. H. de Haan, H. Polinder, and W. L. Kling, "General model for representing variable speed wind turbines in power system dynamics simulations," *IEEE Transactions on Power Systems*, vol. 18, no. 1, February 2003.

18. J. G. Slootweg, S. W. H. de Haan, H. Polinder, and W. L. Kling, "Representing wind turbine electrical generating systems in fundamental frequency simulations," *IEEE Transactions on Energy Conversion*, vol. 18, no. 4, December 2003.

19. Tao Sun, Zhe Chen, and Frede Blaabjerg, "Transient stability of DFIG wind turbines at an external short-circuit fault," *Journal of Wind Energy*, vol. 8, pp. 345–360, 2005.

20. F. Michael Hughes, Olimpo Anaya-Lara, Nicholas Jenkins, and Goran Strbac, "Control of DFIG-based wind generation for power network support," *IEEE Transactions on Power Systems*, vol. 20, no. 4, November 2005.

21. A. Mullane and G. Lightbody, "Wind-turbine fault-ride through enhancement," *IEEE Transactions on Power Systems*, vol. 20, no. 4, November 2005.

22. C. Chompoo-inwai, C. Yingvivatanapong, K. Methaprayoon, and W.-J. Lee, "Reactive compensation techniques to improve the ride-through capability of wind turbine during disturbance," *IEEE Trans. Ind. Appl.*, vol. 41, no. 3, pp. 666–672, May–June 2005.

23. J. Morren and S. W. de Hann, "Ride through of wind turbines with doubly-fed induction generator during a voltage dip," *IEEE Trans. Energy Conversion*, vol. 20, no. 2, pp. 435–441, June 2005.

24. G. Saccomando, J. Svensson, and A. Sannino, "Improving voltage disturbance rejection for variable-speed wind turbines," *IEEE Trans. Energy Conversion*, vol. 17, no. 3, pp. 422–428, September 2002.

25. J. Holtz and W. Lotzkat, "Controlled AC drives with ride-through capability at power interruption," *IEEE Trans. Ind. Appl.*, vol. 30, no. 5, pp. 1275–1283, September–October 1994.

26. W. Qiao, W. Zhou, J. M. Aller, and R. G. Harley, "Wind speed estimation based sensorless output maximization control for a wind turbine driving a DFIG," *IEEE Trans. Power Electronics*, vol. 23, no. 3, pp. 1156–1169, May 2008.

27. T. Sun, Z. Chen, and F. Blaabjerg, "Flicker study on variable speed wind turbines with doubly fed induction generators," *IEEE Trans. Energy Conversion*, vol. 20, no. 4, pp. 896–905, December 2005.
28. G. Lalor, A. Mullane, and M. O'Malley, "Frequency control and wind turbine technologies," *IEEE Trans. Power Systems*, vol. 20, no. 4, pp. 1905–1913, August 2004.
29. J. Ekanayake and N. Jenkins, "Comparison of the response of doubly fed and fixed-speed induction generator wind turbines to changes in network frequency," *IEEE Trans. Energy Conversion*, vol. 19, no. 4, pp. 800–802, December 2004.
30. J. Morren, S. W. H. de Haan, W. L. Kling, and J. A. Ferreira, "Wind turbines emulating inertia and supporting primary frequency control," *IEEE Trans. Power Systems*, vol. 21, no. 1, pp. 433–434, February 2006.
31. G. Ramtharan, J. B. Ekanayake, and N. Jenkins, "Frequency support from doubly fed induction generator wind turbines," *IET Renewable Power Generation*, vol. 1, no. 1, pp. 3–9, March 2007.
32. R. G. de Almeida and J. A. P. Lopes, "Participation of doubly fed induction wind generators in system frequency regulation," *IEEE Trans. Power Systems*, vol. 22, no. 3, pp. 944–950, August 2007.
33. R. G. de Almeida, E. D. Castronuovo, and J. A. P. Lopes, "Optimum generation control in wind parks when carrying out system operator requests," *IEEE Trans. Power Systems*, vol. 21, no. 2, pp. 718–725, May 2006.
34. P. Flannery and G. Venkataramanan, "A fault tolerant doubly fed induction generator wind turbine using a parallel grid side rectifier and series grid side converter," *IEEE Trans. Power Electronics*, vol. 23, no. 3, pp. 1126–1135, May 2008.
35. C. Rahmann, H. -J. Haubrich, L. Vargas, and M. B. C. Salles, "Investigation of DFIG with fault-ride-through capability in weak power systems," [online].
36. S. B. Papaefthimiou and S. A. Papathanassiou, "Simulation and Control of a Variable Speed Wind Turbine with Synchronous Generator," Paper no. 593 [online].
37. T. H. Nguyen and D. -C. Lee, "Ride-Through Technique for PMSG Wind Turbines using Energy Storage Systems," Journal of Power Electronics, Vol. 10, No. 6, November 2010, pp. 733-738.
38. Pedro Lobato, A. Cruz, J. Silva, A. J. Pires, "The Switched Reluctance Generator for Wind Power Conversion," [online].

Index

Printed and bound by CPI Group (UK) Ltd, Croydon, CR0 4YY

18/10/2024

01776267-0003